GEOTECHNICAL SPECI

RECYCLED MATERIALS IN GEOTECHNICAL APPLICATIONS

PROCEEDINGS OF SESSIONS OF GEO-CONGRESS 98

SPONSORED BY
The Soil Properties Committee of
The Geo-Institute of the American Society of Civil Engineers

CO-SPONSORED BY
Center for Innovative Grouting Materials and Technology
(CIGMAT)

October 18–21, 1998
Boston, Massachusetts

EDITED BY
C. Vipulanandan
David J. Elton

ASCE *American Society of Civil Engineers*
1801 ALEXANDER BELL DRIVE
RESTON, VIRGINIA 20191–4400

Abstract: This proceedings, *Recycled Materials in Geotechnical Applications*, contains papers presented at sessions sponsored by the Geo-Institute in conjunction with the ASCE Annual Convention held in Boston, Massachusetts, October 18-21, 1998. They discuss field applications and laboratory testing related to recycled materials. Application-oriented papers were presented on geotechnics of industrial by-products; paper mill sludge for landfill cover; mitigation of void development under bridge approach slabs using rubber tire chips; tire shreds as lightweight fill for embankments and retaining walls; performance of a highway embankment and hydraulic barriers constructed using waste foundry sand; and recycled materials for embankment construction. Papers on laboratory testing of recycled materials included lagoon-stored lime for embankment; construction and demolition debris for base and subbase applications; fly ash for flowable fill, pavement, earth structures and aggregate; repeated loading of stabilized recycled aggregate base course; compaction of contaminated soils-reuse as a road base material; and database on beneficial reuse of foundry by-products.

Library of Congress Cataloging-in-Publication Data

Recycled materials in geotechnical applications: proceedings of sessions sponsored by the Soil Properties Committee of the Geo-Institute of the American Society of Civil Engineers in conjunction with the ASCE National Convention, co-sponsored by the Center for Innovative Grouting Materials and Technology (CIGMAT), Boston, Massachusetts, October 18-21, 1998 / edited by C. Vipulanandan and David J. Elton.

     p.   cm.   –(Geotechnical special publication; no. 79)
    Includes bibliographical references and index.
    ISBN 0-7844-0387-2
    1. Engineering geology–Materials–Congresses. 2. Recycled materials–Congresses. 3. Landfill final covers–Materials–Congresses. 4. Construction and demolition debris–Congresses.
I. Vipulanandan, Cumaraswamy, 1956-   . II. Elton, David J. III. American Society of Civil Engineers. Geo-Institute. Soil Properties Committee. IV. Center for Innovative Grouting Materials and Technology. V. ASCE National Convention (1998: Boston, Mass.) VI. Series.
    TA703.5.R43   1998   98-39145
    624.1'5–dc21   CIP

Geotechnical Special Publications

PREFACE

Billions of tons of recycled materials are being produced annually around the nation. Disposal of these materials is a major problem. But these recycled materials have the potential for use in various geotechnical and geoenvironmental applications. This conference on recycled materials was conceived as a forum review current state-of-the-art in applications, emerging markets for recycled materials and material testing. One objective of this conference and the resulting proceedings is to provide an update on the use of recycled materials and regulatory compliance.

This conference also provides an effective means of sharing recent technological advances, engineering applications and research results among practitioners, researchers and potential users. Selecting and testing recycled materials for various geotechnical applications are issues confronting the profession and will be discussed during the conference.

This proceedings contains some of the papers that were presented at the ASCE Annual Convention held in Boston, Massachusetts, October 18-21, 1998. These sessions were sponsored by the Soil Properties Committee, Geo-Institute, of the ASCE. Papers published in this proceedings were peer reviewed for its content and quality. The standard for the peer review were essentially the same (at least two reviewers) as those for papers being reviewed for possible publication in the ASCE Journal of Geotechnical and Geoenvironmental Engineering. All papers are eligible for discussion in the ASCE Journal of Geotechnical and Geoenvironmental Engineering as well as for ASCE awards.

The Editors would like to thank the paper authors for their cooperation under a tight schedule. Reviewers for this publication were:

Don J. De Groot	P. W. Jayawickrama	Richard A. Reid
Tuncer B. Edil	Joseph F. Labuz	Khaled Sobhan
David J. Elton	David Mast	Vivek Tandon
Khaldoun Fahoum	Jay N. Meegoda	C. Vipulanandan
Patrick Fox	Tarun R. Naik	Daniel O. Wong
Dana Humphrey	Anand J. Puppala	Thomas Yanoshak
Hilary I. Inyang	Krishna R. Reddy	

Reviewers constructive and timely reviews are very much appreciated.

Editors
C.Vipulanandan (Vipu)
Center for Innovative Grouting Materials and Technology (CIGMAT)
University of Houston

and

David J. Elton
Auburn University

Contents

GEOTECHNICS OF INDUSTRIAL BY-PRODUCTS

Tuncer B. Edil[1] and Craig H. Benson[2], Members ASCE

ABSTRACT

Large volumes of industrial by-products are generated each year by U.S. industries, and most are landfilled as solid waste at considerable expense. However, many by-products have desirable properties and leach contaminants at such low concentrations that their use in construction applications does not present an environmental hazard. As a result, environmental regulations are being modified to permit re-use of these materials in a variety of applications. Ideal applications for these materials exist in the transportation, construction, and environmental industries, where large volumes of earthen materials are used each year. This paper focuses on the principles and protocols to be followed in developing the geotechnics of industrial by-products and how to deal with their unusual characteristics. These issues are discussed with examples derived from the writers' experience.

INTRODUCTION

U.S. industries annually generate millions of metric tons of solid by-products (Miller and Collins 1976). Most of these materials have been landfilled at considerable cost since the inception of modern environmental regulations in the late 1970s and early 1980s. Recently there has been a shift in societal attitudes resulting in strong interest in developing beneficial re-use markets for industrial by-products. As a result, environmental regulations have changed and beneficial re-use of industrial by-products is now permissible in a variety of applications. In fact, blast furnace slag, fly ash, bottom ash, boiler slag, reclaimed pavement materials, anthratic coal waste, and many other industrial by-products have been or

[1] Professor, Dept. of Civil and Environmental Engineering, University of Wisconsin-Madison, Madison, WI 53706, edil@engr.wisc.edu, Ph. (608)262-3225

[2] Associate Professor, Dept. of Civil and Environmental Engineering, University of Wisconsin-Madison, Madison, WI 53706, chbenson@facstaff.wisc.edu, Ph. (608)262-7242

are in the process of being beneficially used as highway construction materials (Miller and Collins, 1976).

The transportation, construction, and environmental industries have the greatest potential for re-use because they use vast quantities of earthen materials annually.   Replacement of natural soils, aggregates, and cements with solid industrial by-products is highly desirable. In some cases, a by-product is inferior to traditional earthen materials, but its lower cost makes it an attractive alternative if adequate performance can be obtained.   In other cases, a by-product may have attributes superior to those of traditional earthen materials. Often select materials are added to industrial by-products to generate a material with well-controlled and superior properties.

This paper focuses on the principles and protocols to be followed in developing the geotechnics of industrial by-products and how to deal with their unusual characteristics. These issues are discussed with examples derived from the writers' experience.

DEVELOPMENT OF INDUSTRIAL BY-PRODUCTS AS GEO-MATERIALS

Specific issues that relate to the development of industrial by-products as geo-materials include:

1. Identification of the application
2. Selection of the key properties required for the application
3. Environmental suitability
4. Laboratory testing protocols
5. Modeling of engineering behavior
6. Constructability and field verification of performance
7. Construction specifications
8. Long-term performance
9. Dissemination of technical information

IDENTIFICATION OF APPLICATION

Identification of appropriate geo-applications for an industrial by-product is the most crucial step. This step requires consideration of the salient properties of the by-product and a comprehensive knowledge and understanding of geotechnical construction, economics, and environmental regulations. Matching the beneficial attributes of a by-product with the requirements of various applications is essentially an entrepreneurial activity that often evolves through interactions and communications between individuals with different technical backgrounds. Innovation is a key characteristic of this process.   The following example applications illustrate this point.

Shredded Tires

Tire products contain carbon-based materials, which are known to be highly sorbent materials (Miller and Chadik 1993). Environmental spills of toxic chemicals and contaminated groundwater are critical issues facing the remediation industry. Scrap tires, produced at a rate of over 240 million per year (Scrap Tire Management Council 1990, U.S. EPA, 1991), can be used in applications that require inexpensive sorbents. Shredded scrap tires can be used above landfill liners as a leachate collection layer (Hall 1991), and as a reactive layer for leachate treatment or in landfill covers as a gas control layer (Edil et al. 1996, Park et al. 1996a, Kim et al. 1997). Ground scrap tires can also be incorporated into slurry walls to enhance their ability to contain organic chemicals (Park et al. 1996b, 1997).

Shredded tires exhibit excellent frictional properties. Therefore, they can be used to enhance the strength properties of soils by internal fiber reinforcement. In addition, because of their lower specific gravity (1.15 to 1.21), relative to that of soil solids (2.55 to 2.75), tire chips, alone or in mixtures with soils, offer an excellent light-weight and strong fill material for use in fills, earthen structures, etc. (Humphrey et al. 1992, Upton and Machan 1993, Humphrey et al. 1993, Ahmed and Lovell 1993, Edil and Bosscher 1994, Foose et al. 1996, Tatlisoz et al. 1997). They also can be used in embankments over soft ground or as backfill behind retaining structures (Read et al. 1991, Bosscher et al. 1992, Tatlisoz et al. 1998).

Foundry Sands

Excess foundry sands from metal casting plants often contain a significant amount of bentonite (up to 15%), and thus can have low hydraulic conductivity. Consisting of primarily inert sand minerals, they also show low sensitivity to compaction moisture and effort and to environmental impacts such as wet and dry and freeze and thaw exposures. These attributes make some foundry sands excellent hydraulic barrier materials (Stephen and Kunes 1982, Freber 1996, Vierbicher Associates 1996, Abichou et al. 1998a).

Excess foundry sands, especially the ones containing lower amounts of bentonite, i.e. less than 5 to 10 %, exhibit excellent mechanical properties and can be used as a substitute for select sand aggregate used in constructing highway subbases and asphalt or Portland cement concrete in pavement structures (AFS 1991, AFS 1992, AFS 1996, Javed and Lovell 1994, Javed 1994, McIntyre et al. 1991, Mast 1997, Kleven et al. 1998, Abichou et al. 1998b).

Paper Mill Sludge

Residues from wastewater treatment plants at paper mills, called paper mill sludges, have been used to construct hydraulic barrier layers in landfill covers (Kraus et al. 1997). These sludges are comprised of organic material and mineral fines (typically kaolinite and calcite) and can have low hydraulic conductivity when compacted properly. In addition, paper mill sludges are mixed with sands and clayey soils to make topsoil.

Sludges have also been processed by firing to produce a lightweight high strength vitrified aggregate. For example, Minergy Corporation operates a sludge processing plant near Neenah, Wisconsin. Fiber in the sludge is used to fuel the firing operation to vitrify the clay minerals. Also, excess heat from firing is used to produce steam for an adjacent paper mill (Boudwin 1997).

Plastics

Shredded plastics have been used to reinforce soils used for structural fill. For example, Benson and Khire (1994) found that the shear strength and stiffness of sand can be enhanced significantly by mixing the sand with small strips of scrap high density polyethylene. An advantage of this re-use application is that commingled plastics can be used, whereas re-melting and molding plastic is precluded when waste plastics are commingled. In addition, using plastic as soil reinforcement requires little processing or energy use beyond that used in shredding.

SELECTION OF KEY PROPERTIES

Each application requires a set of key properties that need to be evaluated for assessing the suitability of a given by-product. Properties related to design are needed, along with those important to constructability and environmental suitability. There are primary properties as well as secondary properties. For instance, when using tire chips as construction fill material, the primary properties are strength, deformability, and constructability (Bosscher et al. 1992). The secondary properties include long-term compression, combustibility, and environmental suitability. If the fill is for a highway embankment, resilient modulus and plastic strain become the primary properties to be evaluated (Bosscher et al. 1997). If the fill is for a geosynthetic-reinforced backfill for a retaining structure, in addition to strength and compressibility, the primary properties needed are the interaction coefficient and pullout resistance of the geosynthetic in the tire chips fill (Tatlisoz et al. 1998).

If foundry sands are being considered as aggregate for a highway subbase or pavement (asphalt or Portland cement concrete), the primary properties are similarly resilient modulus and plastic strain. The secondary properties are ability to drain (i.e. hydraulic conductivity) and resistance to frost action. For use in a hydraulic barrier, the primary property to be evaluated is hydraulic conductivity. However, physical (frost and desiccation) and chemical durability of hydraulic conductivity are just as important.

Constructability is common to all of these geotechnical applications, but standard tests to evaluate constructability do not exist. Compaction moisture, type, and effort can be evaluated in the laboratory and optimum compaction conditions can be identified. But, material handling and workability typically require field demonstration and trials for optimization. For instance, fly ash, either as Class C (cementitious) or as a mixture of Class C and Class F (non reactive) can be mixed with sand or bottom ash and water to generate a reactive hydraulic barrier with low hydraulic conductivity when compacted properly (Bowders et al. 1987, Edil et al. 1987, Palmer et al. 1995). However, one of the greatest challenges of using these materials is proper mixing and handling of the mix before it sets up in the field (Palmer et al. 1995).

ENVIRONMENTAL SUITABILITY

Until recently industrial by-products have been considered solid or hazardous waste and typically have been landfilled. During the last five years, however, environmental agencies have become aware of the beneficial attributes of many by-products and the high cost of landfilling them. As a result, re-use of industrial byproducts is now encouraged. Unlike natural earthen materials, however, the potential for pollution by by-products has to be assessed in the context and environment of a given application. Unfortunately no standard method currently exists to make such an assessment.

Environmental suitability assessment first involves determining if a by-product is a hazardous material. This is typically accomplished by performing a Toxicity Characterization Leachability Procedure (TCLP) (EPA 1986), which has replaced the older EP-Toxicity test (Extraction Procedure Toxicity Test). Once a material has been found non-hazardous, additional batch tests are performed to determine equilibrium concentrations of chemicals leached from the waste material. Container volumes and weight of the waste may vary, but the liquid is usually water to simulate the leaching condition likely to exist in the field. For example, the Wisconsin Administrative code categorizes by-products based on concentrations measured in the ASTM D 3987 water leach test. Concentrations obtained from a leach test and limits set by various standards [(i.e., maximum contaminant levels (MCLs) and protective action levels (PALs)] have to be converted from the mass of compound per volume of liquid to the mass of compound per mass of waste. Only then, can fair comparisons be made between results of a leaching test and regulatory standards.

Column tests simulating field conditions are also performed using liquids to ascertain whether water quality will be affected. The aquatic environment in which the industrial by-product will be used must be carefully simulated if the results are to be applicable to the field. The following factors are typically considered: pH and ionic strength of the aquatic environment, presence of humic acids, immersion time, stagnant vs. flowing conditions, and microbial activity.

Tire Chips

An extensive literature survey by Tatlisoz et al. (1996) regarding the environmental suitability of tire chips indicates that concentrations of inorganic or organic compounds exceeding maximum contaminant limits have not been found in leaching studies. Nevertheless, some inorganic constituents (e.g., zinc, iron, barium, manganese) have been found in concentrations higher than drinking water standards. However, many of the tests were designed to simulate "worst case" conditions and do not simulate typical environments in which tire chips are used. Thus, lower concentrations are likely to exist in the field. To demonstrate environmental suitability in the field, basin lysimeters can be placed under the industrial by-product from which water samples are collected to assess environmental suitability. An example application of basin lysimeters is described in a subsequent section.

Foundry Sands

Ham et al. (1981) conducted a laboratory and field study on the leaching potential of foundry wastes. The laboratory study utilized the TCLP and the American Foundrymen's Society (AFS) test, which was developed for leach testing of foundry wastes. Lovejoy et al. (1996) report that none of the foundry sands tested using the TCLP test classified as hazardous according to the Resource Conservation and Recovery Act (RCRA) definition. However, iron concentrations from both tests exceeded drinking water standards. Hardness, barium, copper, and zinc concentrations were lower than those obtained from leach testing of typical Wisconsin soils, whereas the fluoride concentration and pH were significantly higher. Ground water quality data collected during the field study showed no impacts that can be attributed to the foundry sand or the natural soils used in the study.

LABORATORY TESTING PROTOCOLS

ASTM standards are the main source for measuring various properties of soil and rock. These consensus standards have been developed after many years of testing and are based not only on the proper understanding of a property and its testing but also on the relevance of that property to design and construction. There has been a welcome proliferation of ASTM soil and rock testing standards in recent years; however, nearly all of these standards have been developed for naturally occurring earthen materials. Therefore, their direct application to solid industrial by-products is unwarranted without an investigation of applicability.

For instance, compaction standards (ASTM D 698 and D 1557) have been written for earthen materials with a maximum particle size of 19 mm. Application of these compaction standards to tire chips, which typically have unequal dimensions in the range of 25 to 300 mm, is not appropriate. A larger compaction mold (305 mm in diameter) with a larger compaction shoe has to be used. Edil and

Bosscher (1994) assessed how various testing factors affected compaction characteristics of tire chip and tire chips-soil mixtures. Results of their study show that a hammer weight of 272 N, hammer drop of 0.45 m, and 14 or 64 blows per layer (depending on the desired effort) are suitable for compaction of tire chips and soil-tire chip mixtures.

Another example is conducting Atterberg limits tests, hydrometer analysis, and compaction of foundry sands. Foundry sands contain highly dehydrated bentonite. The relevant ASTM standards (D 4318, D 422, D 698, and D 1557) call for standing times (hydration periods) up to 16 to 18 hours for soils of similar character prior to testing. While this is a reasonable hydration period for most soils, it is inadequate for foundry sands. Kleven et al. (1998) show that the Atterberg limits change as a function of hydration time (Fig. 1). A week of hydration was found to be adequate for hydration of most foundry sands.

MODELING ENGINEERING BEHAVIOR

Like natural soils and rocks, the engineering behavior of industrial by-products needs to be represented by appropriate mathematical models for analysis and design. Many years of experience exist in developing models for natural earthen materials that describe mechanical behavior and conduction phenomena. Existing soil models can be a starting point for describing the behavior of industrial products. However, they have to be verified to ensure that the behavior of the by-product is consistent with the behavior predicted by the model.

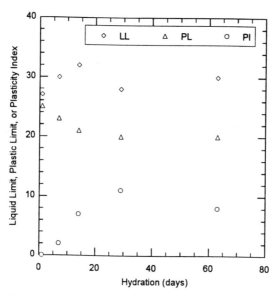

Fig. 1. Atterberg Limits of Foundry Sand as a Function of Hydration Time (adapted from Kleven et al. 1998).

Mixtures of tire chips and sand exhibit non-linear plastic deformation in laterally constrained one-dimensional compression, as shown in Fig. 2 (Edil and Bosscher 1994, Tatlisoz et al. 1997). However, after the initial cycle of loading, the behavior is essentially linear and predominantly elastic if the stress level is maintained above a certain level (e.g., > 40 kPa). Therefore, for repetitive loading far below the failure stress (e.g., highway traffic loading), the material behavior can be approximated with an elastic model. Such modeling proved to be successful in describing the response of tire chips and tire chip-sand mixtures in large-scale physical simulations and field demonstrations of a test embankment (Bosscher et al. 1997). Alternatively a log-log relationship can be used to describe a non-linear modulus covering the whole range of stresses.

The shear strength of mixtures of loose sand or sandy silt and tire chips can be represented by a linear Mohr-Coulomb failure criterion (Tatlisoz et al. 1997). However, mixtures of tire chips with dense sand require a bilinear failure envelope (Foose et al. 1996) as shown in Fig. 3. In dense sand, greater interaction between the tire chips and sand matrix occurs. Bilinear failure criteria have also been used to describe the shear strength of sands mixed with other reinforcing materials (Gray and Ohashi 1983).

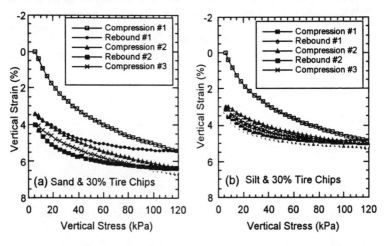

Fig. 2. Stress-Strain Curves from Laterally Constrained Compression Tests: (a) Sand and 30% Tire Chips and (b) Sandy Silt and 30% Tire Chips.

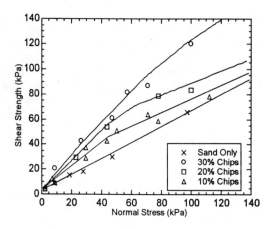

Fig. 3. Shear Strength Envelopes for Dense Portage Sand-Tire Chip Mixtures (adapted from Foose et al. 1996)

Modeling foundry sands is more straightforward since they consist of earthen minerals (sand and bentonite), albeit in proportions not normally found in nature. The resilient behavior of foundry sands used as subbase material can be represented using models developed for soils (Kleven et al. 1998). Similar statements can be made regarding the hydraulic properties of foundry sands, as well as the hydraulic properties of paper sludge and compacted fly ash. For example, paper sludge can be treated in the same manner as compressible soft clay or organic soil (Kraus et al. 1997). However, a more in-depth understanding of the behavior of these materials can require a different approach. For example, network percolation models are being developed to simulate flow through foundry sands and should provide a more fundamental understanding of the function of the bentonite in the pore space. Such a model may also lead to a better understanding of the behavior of soil-bentonite mixtures commonly used in geotechnical practice.

VERIFICATION OF CONSTRUCTABILITY AND FIELD PERFORMANCE

Geotechnical engineering is a practice that has evolved through field observations. Laboratory tests, large-scale models, and theoretically based models of behavior are all tools to predict field performance. Due to inherent variability of geo-materials and inevitable variations from the simplifying assumptions used in analysis, field verification is essential to verify whether models capture the field response. It is also needed to assess constructability. In a similar vein, field performance verification (in some cases in the long-term) is essential for widespread acceptance of industrial by-products by the engineering community. For example, field demonstrations consisting of test embankments and monitoring of constructed facilities (Bosscher et al. 1992, Humphrey et al. 1993) have been necessary for tire chips to be accepted as a fill material.

Advanced planning, clear objectives, adequate data collection, and in-depth interpretations are required if field demonstrations are to be useful. To have these elements incorporated in an on-going project is harder than in specific demonstration projects. All too often field demonstrations are conducted without careful planning, and the resulting data have little value. Constructing a field demonstration also helps delineate construction and material handling problems. And, with some planning, the construction phase can be turned into an experiment to identify optimum construction techniques.

Tire Chips

Several elements were introduced into the construction program for a test embankment using tire chips provided major insights regarding constructability (Bosscher et al. 1992). For instance, the embankment was constructed in eight sections including two control sections to permit testing of different configurations. Some sections had only tire chips while others had a mixture of tire chips and sand or layers of tire chips and sand (Fig. 4). This allowed a comparison of ease of construction and the settlement of different configurations. Compacting one side with a vibratory compactor and the other with static compaction showed that vibration improved compaction slightly, but not enough to support the use of vibratory compactors. Similarly, systematic measurements of density revealed that there was no significant increase in density with more than one pass of the roller.

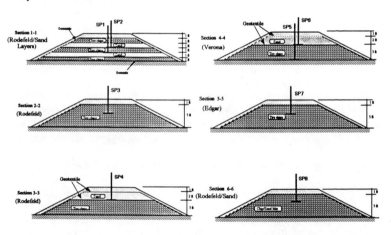

Fig. 4. Sections of Test Embankment Constructed by Bosscher et al. (1992).

Three different sizes of tire chips were used to assess the influence of chip size on material handling and post-construction performance. Until this field test, chip size was considered very important and strictly specified in construction specifications. The test embankment showed that chips size is not an important parameter. Very large shreds (up to 300 mm in length) produced with one cycle of shredding, though somewhat more difficult to handle during construction, provided comparable or less settlement than smaller chips (about 50 to 75 mm) under self weight.

To evaluate fill performance under traffic loading, rutting and pavement quality monitoring data were collected after truck traffic loading. Predictions were then made of damage accumulation using a layered pavement structural analysis based on elastic theory (Bosscher et al. 1997). Properties of the tire chip fill measured in the laboratory were used in the elastic analysis. The measured damage in the field was consistent with the predicted damage to the pavement.

Leachate quality of tire fills, based on laboratory leach tests, was questioned heavily. As mentioned earlier, laboratory leach test results are often difficult to extrapolate to the field, even though they are useful to identify clear-cut cases. Collection basin lysimeters constructed at the bottom of the test embankment allowed monitoring of the quality of tire fill leachate under field conditions. The data indicated little likelihood of shredded tires having adverse effects on groundwater quality and helped improve the regulatory views of re-use of tires in construction.

Fly Ash

Several studies were performed to investigate the use of coal fly ash in hydraulic barrier construction (Sachdev and Amdurer 1985, Edil et al. 1987, Bowders et al. 1987, Bowders et al. 1990, Creek and Shackelford 1992, Shakleford and Glade 1994, Palmer et al. 1995). Palmer et al. (1995) conducted a field demonstration of hydraulic barrier construction using fly ash. They show that construction with pozzolanic fly ash can be challenging because of the rapid set time of Class C fly ashes. The construction method must be rapid and efficient so that a relatively uniform layer of fly ash can be placed and compacted before the material cures and becomes unmanageable. Also, cold joints are a serious problem with fly ash barriers. They can be the primary flow paths and govern the hydraulic conductivity of the barrier if not properly sealed.

Other Materials

Field demonstrations with foundry sands and paper mill sludge have shown that these materials can be handled like soils. However, as with soils, heterogeneity can be a significant problem and the material must be carefully inspected when delivered. For example, Abichou et al. (1998a) found that the foundry sand delivered for a field demonstration had much lower bentonite content than the foundry sand used to develop the construction specifications. As a result, additional laboratory testing was needed to revise the specifications to meet the project objectives. In addition, materials delivered to the job site often contain objects larger than the buckets usually used to deliver materials to the laboratory. These materials often need to be removed or the specifications need to be modified to ensure that these larger objects do not affect the field performance.

CONSTRUCTION SPECIFICATIONS

Construction specifications are difficult to generate for industrial by-products because of the lack of extensive experience with such materials. General principles can be derived from earth construction specifications and have to be customized for by-products based on their observed behavior during laboratory compaction tests and field constructability observations. Research on behavior of by-products as geo-materials should include recommendations for construction specifications to guide engineers. An example of recommendations for testing and construction for tire chips construction is given in Edil and Bosscher (1994).

Construction specifications can be readily developed for some materials. For example, specifications for clay liners can be adopted for constructing barrier layers with foundry sands. Acceptance zones for compaction can be developed using common methods (Daniel and Benson, 1990) and are often less restrictive than those for clay liners. Lighter compaction equipment can be employed (e.g., 15,000 kg), and compactors with shorter feet can be used (e.g., padfoot compactors with 100 mm feet). Material testing is also similar to clay liners, except that bentonite content should be measured along with the Atterberg limits and particle size distribution. Abichou et al. (1998a) recommend that specifications require that the liquid limit exceed 20 and the bentonite content exceed 6%.

DISSEMINATION OF TECHNICAL INFORMATION

Dissemination of technical information generated from research and case histories involving use of industrial by-products as geo-materials is crucial for widespread adoption of an industrial by-product by the technical community. Unfamiliarity with the material and lack of design and construction guidelines inhibit use of these materials. Passive availability of technical information in various publications does not solve this problem. Information has to be synthesized and formatted in a manner that can be used directly by the engineer. A good example in this regard is how technical information on geosynthetics has resulted in their widespread adoption as geo-materials. With the backing of the geosynthetics manufacturing industry, tools necessary for design and construction have been made available to practicing engineers in industry and government, who set the pace in adoption of innovations.

Beneficial uses of foundry sands have been known for some time, but their use in construction has been limited. This situation is partly a result of the regulatory environment, but it is also attributable to poor dissemination of technical information. Being a major foundry state, Wisconsin is in the process of changing this situation. The Wisconsin Cast Metal Association with the assistance of Wisconsin Industrial By-Products Market Development Board is developing an easily accessible technical database system (Abichou et al. 1998b). This database system is designed to provide a compendium of existing technical information, critical reviews of this information, and delineation of information gaps that preclude development of various applications. The database is available through a web page providing easy access. Similar activities are needed for other by-products.

Wisconsin has also invested significant funds into developing strategies for re-use of its millions of scrap tires through a grant program funded by tire disposal fees. This program generated a large quantity of technical information regarding use of scrap tires as geo-materials. Not having a concerned industry behind it, dissemination of this information has been limited to the efforts of the investigators in publishing the materials and providing collections of papers to the technical community (Use of Scrap Tires in Civil and Environmental Construction, 1996). However, the information is not simplified for direct use beyond the technical information level.

REGULATORY ENVIRONMENT FOR USE OF INDUSTRIAL BY-PRODUCTS

Regulatory hurdles present one of the biggest stumbling blocks for beneficial re-use. Usually a well-defined pathway for regulatory approval does not exist, which impedes the progress of a project, and often increases its cost. Fortunately, many states are now developing codes that provide clear instructions regarding how and when industrial by-products can be used. For example, the Wisconsin Administrative Code now includes Section NR 538 on industrial by-products. Materials are categorized according to their environmental suitability, which is defined by concentrations of contaminants of concern in leachate from a water leach test (ASTM D 3987). Once categorized, a by-product has blanket approval for use in project types listed in the category. Testing is then required on an annual basis to ensure the leaching characteristics of the by-product have not changed.

When a regulatory approval process does not exist, meeting with officials from the regulatory agency is the first step in a beneficial re-use project. Far too often, engineers design most of a project before meeting with the regulatory agency, only to find that the project must be modified significantly to meet regulatory approval. During the first meeting, the agency can be carefully queried to determine what steps are necessary to obtain approval, and what type of data must be generated. Additonal meetings should be held as the project progresses to keep regulatory officials current on the data and the elements of the project. When this approach is employed, regulatory approval often occurs more quickly, and costs are easier to anticipate.

SUMMARY

U.S. industries produce millions of metric tons of by-products annually. Only small portions of these materials are used beneficially; most are landfilled as solid waste. Large quantities of non-hazardous industrial by-products can be used beneficially as geo-materials in civil construction. Such recycling will save millions of dollars annually to U.S. industries in avoided landfill costs, generate cost-effective alternatives to traditional aggregates, minimize environmental damage due to aggregate mining, and provide engineers with new construction materials sometimes with superior qualities. Expansion of beneficial use of industrial by-products requires a well-planned and deliberate development plan. The elements of such a plan have been presented in this paper with examples.

Future studies of beneficial re-use should include a field demonstration element whenever possible. Demonstrations validate that properties anticipated based on laboratory work can be achieved at full-scale in the field. In addition, field demonstrations can be used to optimize construction methods, and often illuminate potential construction problems that are not considered in laboratory studies. Finally, results of beneficial re-use studies need to be disseminated to practicing engineers. Technical publications in engineering journals must be supplemented with design guidance documents that are readily available to practicing engineers in industry and government. The worldwide web is an ideal mechanism to make this type of information widely accessible at minimum cost.

ACKNOWLEDGEMENT

The ideas expressed in this paper have been developed over the years through the writers' involvement in a series of projects funded by a number of agencies. Support was provided by the Wisconsin Dept. of Natural Resources Waste Tire Management or Recovery Grant Program, the Wisconsin Dept. of Transportation, the Univ. of Wisconsin System Groundwater Research Advisory Council, the Univ. of Wisconsin System Solid Waste Research Program, the Wisconsin Industrial By-Product Recycling Market Development Board, the U.S. Environmental Protection Agency, the Wisconsin Cast Metals Association, the Wisconsin Power & Light Corp., and the Wisconsin Electric Power Co. Numerous graduate students have contributed to these projects. In particular, the contributions of our colleagues Drs. Peter J. Bosscher and Jae K. Park are acknowledged.

REFERENCES

Abichou, T., Benson, C., and Edil, T. (1998a), "Beneficial Reuse of Foundry Sands in Construction of Hydraulic Barrier Layers," Environmental Geotechnics Report 98-2, Dept. of Civil and Environmental Engineering, University of Wisconsin-Madison.

Abichou, T., Benson, C., and Edil, T. (1998b), "Database on Beneficial Reuse of Foundry By-Products," Environmental Geotechnics Report 98-3, Dept. of Civil and Environmental Engineering, Univ. of Wisconsin-Madison.

Ahmed, I., and Lovell, C. (1993), "Use of Rubber Tires in Highway Construction." *Utilization of Waste Materials in Civil Engineering Construction*, ASCE, New York, N.Y., 166-181.

American Foundrymen's Society, Inc. (1991), "Final (Phase I) Report on Alternate Utilization of Foundry Waste Sand." Illinois Department of Commerce and Community Affairs. Grant No. 90-82109.

American Foundrymen's Society, Inc. (1992), "Final (Phase II) Report on Alternate Utilization of Foundry Waste Sand." Illinois Department of Commerce and Community Affairs. Grant No. 90-82109.

American Foundrymen's Society, Inc. (1996), "Foundry Sand Beneficial Reuse Manual (Special Report)." Illinois Department of Commerce and Community Affairs. Grant No. 90-82109.

Benson, C., and Khire, M. (1994), "Soil Reinforcement with Strips of Reclaimed HDPE," *J. of Geotech. Eng.*, ASCE, Vol. 120, No. 5, pp. 838-855.

Bosscher, P., Edil, T., and Eldin, N. (1992), "Construction and Performance of a Shredded Waste Tire Test Embankment," *Transportation Research Record*, No. 1345, pp. 44-52.

Bosscher, P., Edil, T., and Kuraoka, S. (1997), "Design of Highway Embankments Using Tire Chips," *J. of Geotech. and Geoenv. Eng.*, ASCE, Vol. 123, No. 4, pp. 295-304.

Boudwin, T. (1997), Personal communication with Thomas Boudwin, Environmental Engineer, Minergy Corporation, Neenah, Wisconsin.

Bowders, J., Usmen, M., and Gidley, J. (1987), "Stabilized Fly Ash for Use as Low-Permeability Barriers," *Geotechnical Practice for Waste Disposal '87*, Geotechnical Special Publication No. 13, ASCE, New York, N.Y., PP. 320-333.

Bowders, J., Gidley, J., and Usmen, M. (1990), "Permeability and Leachate Characteristics of Stabilized Class F Fly Ash," *Transportation Research Board 1288*, TRB, National Research Council, Washington, DC, PP. 70-77.

Creek, D., and Shackleford, C. (1992), "Permeability and Leaching Characteristics of Fly Ash Liner Materials," *Transportation Research Board 1345*, TRB, National Research Council, Washington, DC, PP. 74-83.

Daniel, D. and Benson, C. (1990), "Water Content-Density Criteria for Compacted Soil Liners," *J. of Geotech. Eng.*, ASCE, 116(12), 1811-1830.

Edil, T., Berthouex, P., and Vesperman, K. (1987), "Fly Ash as a Potential Waste Liner," *Geotech. Practice for Waste Disposal '87*, Geotech. Special Publication No. 13, ASCE, pp. 447-461.

Edil, T. and Bosscher, P. (1994) "Engineering Properties of Waste Tire Chips and Soil Mixtures," *Geotech. Testing J.*, ASTM, Vol. 17, No. 4, pp. 453-464.

Edil, T., Kim, J., and Park, J. (1996), "Reactive Barriers for Containment of Organic Compounds," *Proc. 3rd International Symposium on Environmental Geotechnology*, San Diego, California, pp. 523-532.

Foose, G., Benson, C., and Bosscher, P. (1996), "Sand Reinforced with Shredded Waste Tires," *J. of Geotech. Eng.*, ASCE, Vol. 122, No. 9, 760-767.

Freber, B. W. (1996), "Beneficial Reuse of Selected Foundry Waste Material," *Proceeding of 19th International Madison Waste Conference*, Madison, WI, No. 13, Sept. 1996, pp. 246-257.

Gray, D. and Ohashi, H. (1983), "Mechanics of Fiber Reinforcement in Sand," *J. of Geotech. Eng.*, ASCE, Vol. 109, No. 3, pp. 335-353.

Ham, R., Boyle, C., and Kunes, T. (1981), "Leachability of Foundry Process Solid Wastes," *J. Environmental Eng.*, ASCE, 107, No. 1, pp. 155-170.

Hall, T., (1991), "Reuse of Shredded Tire Material for Leachate Collection Systems." *Proceedings 14th Annual Madison Waste Conference*, Dept. Engineering Professional Development, University of Wisconsin, Madison, WI., 367-376.

Humphrey, D., and Manion, W. (1992), "Properties of Tire Chips for Light-Weight Fill," *Grouting, Soil Improvement, and Geosynthetics*, Vol. 2, ASCE, New York, N.Y., 1344-1355.

Humphrey, D., Sandford, T., Cribbs, M., and Manion, W. (1993), "Shear Strength and Compressibility of Tire Chips for Use as Retaining Wall Backfill," *Transportation Research Record*, No. 1422, Transportation Research Board, Washington DC, pp. 29-35.

Javed, S. (1994), "Use of Waste Foundry Sand in Highway Construction," Report JHRP-94/2, School of Civil Engineering, Purdue Univ., West Lafayette, IN.

Javed S., and Lovell, C. (1994), "Use of Waste Foundry Sand in Civil Engineering." *Transportation Research Record*, No. 1486, Transportation Research Board, Washington D.C., 1995, pp. 109-113.

Kraus, J., Benson, C., Maltby, V., and Wang, X. (1997), "Field and Laboratory Hydraulic Conductivity of Compacted Papermill Sludges," *J. of Geotech. and Geoenv. Eng.*, ASCE, Vol. 123, No. 7, pp. 654-662.

Kim, J., Park, J., and Edil, T. (1997), "Sorption of Organic Compounds in the Aqueous Phase onto Tire Rubber," *J. of Environmental Eng.*, ASCE, Vol. 123, No. 9, pp. 827-835.

Kim, J., Park, J., Edil, T., and Jhung, J. (1996), "Sorption Capacity of Ground Tires for Vapor Phase Volatile Organic Compounds," *Volatile Organic Compounds in the Environment, ASTM STP 1261*, W. Wang, J. Schnoor, and J. Doi, Eds., pp. 237-244.

Kleven, J., Edil, T., and Benson, C. (1998), "Beneficial Re-use of Foundry Sands in Roadway Subbase," Environmental Geotechnics Report 98-1, Dept. of Civil and Environmental Engineering, University of Wisconsin-Madison.

Lovejoy, M., Ham, R., Traeger, P., Wellander, D., Hippe, J., and Boyle W. (1996), "Evaluation of Selected Foundry Wastes for Use in Highway Construction," *Nineteenth Annual Madison Waste Conference*, University of Wisconsin-Madison, Wisconsin, pp. 19-31.

Mast G. (1997), "Field Demonstration of a Highway Embankment Using Waste Foundry Sand," M. Sc. Thesis, Purdue University. Purdue, IN.

McIntyre, S., Rundman, K., Bailhood, C. , Rush, P., Sandell, J., and Stillwell, B. (1991), "The Benefication and Reuse of Foundry Sand Residuals," Michigan Technological University, Houghton, Michigan.

Miller, W., and Chadik, P. (1993) "A Study of Waste Tire Leachability in Potential Disposal and Usage Environments," Amended Final Report to Florida Department of Environmental Regulation No. SW67.

Miller R., and Collins R. (1976) "Waste Materials as Potential Replacement for Highway Aggregates". National Cooperative Highway Research Program report No. 166, *Transportation Research Board*, 1976, pp. 1-24.

Palmer, B., Benson, C., and Edil, T. (1995), "Hydraulic Characteristics of Class F Fly Ash as a Barrier Material: Laboratory and Field Evaluation," Environmental Geotechnics Report 95-8, Dept. of Civil and Environmental Engineering, University of Wisconsin-Madison.

Park, J., Kim, J., and Edil, T. (1996a), "Mitigation of Organic Compound Movement in Landfills by a Layer of Shredded Tires," *Water Environment Research.*, Vol. 68, No. 1, pp. 4-10.

Park, J., Kim, J., Edil, T., and Madsen, C. (1996b), "Use of Ground Tires for Organic Compound Containment in the Soil-Bentonite (SB) Slurry Cutoff Wall," *IS-Osaka '96*, 2nd International Congress on Environmental Geotechnics, Osaka, Japan, Vol. 2, pp. 881-886.

Park, J., Kim, J., Madsen, C., and Edil, T. (1997), "Retardation of Volatile Organic Compound Movement by a Soil-Bentonite Slurry Cutoff Wall Amended with Ground Tire," *Water Environment Research*, Vol. 69, No. 5, pp. 1022-1031.

Read, J., Dodson, T., and Thomas, J. (1991), "Use of Shredded Tires for Lightweight Fill," Report No. DTFH-71-90501-OR-11, Oregon Department of Transportation, Highway Division, Rd. Sec., Salem, Oregon.

Sachdev, D., and Amdurer, M. (1985), "Fly Ash a Waste Pile Liner Material: Feasibility Study," Envirosphere Co., New York, N.Y.

Scrap Tire Management Council (1990), "Scrap Tire Use/Disposal Study," 1-4

Shackleford, C., and Glade, M. (1994), "Constant-Flow and Constant-Gradient Permeability on Sand-Bentonite-Fly Ash Mixtures," *Hydraulic Conductivity and Waste Contaminant Transport in Soil, ASTM STP 1142*, David Daniel and Stephen J. Trauwein, Eds., ASTM, Philadelphia, PA, PP. 184-223.

Tatlisoz, N., Benson, C., and Edil, T. (1997), "Effect of Fines on Mechanical Properties of Soil-Tire Chip Mixtures," *Testing Soil Mixed with Waste or Recycled Materials, ASTM STP 1275*, M. A. Wasemiller and K. B. Hoddinott, Eds., pp. 93-108.

Tatlisoz, N., Edil, T., and Benson, C. (1998), "Geosynthetic Reinforced Walls and Embankments Using Soil-Tire Chip Mixtures," *J. of Geotech. and Geoenv. Eng.*, ASCE, (submitted for publication).

Tatlisoz, N., Edil, T., Benson, C., Park, J., and Kim, J. (1996), "Review of Environmental Suitability of Scrap Tires," Environmental Geotechnics Report 96-7, Dept. of Civil and Environmental Engineering, University of Wisconsin-Madison.

Stephen, W., and Kunes P. (1982), "Cutting the Cost of Disposal Through Innovative and Constructive Uses of Foundry Wastes," *AFS Transactions*, Vol. 81-84, pp. 697-708.

Upton, R., and Machan, G. (1993) "Use of Shredded Tires for Lightweight fill." *Transportation Research Records*, No. 1422, Transportation Research Board, Washington, D.C., 36-45.

U.S. EPA (1991) "Market for Scrap Tires," EPA/530-SW-90-074B, U.S. EPA.

Use of Scrap Tires in Civil and Environmental Construction (1996), Selected Publications from the University of Wisconsin-Madison, Environmental Geotechnics Report No. 96-3, Madison, WI.

Vierbicher Associates (1996), "Final Report: Beneficial Reuse of Selected Foundry Waste Material," Prepared for Wisconsin Dept. of Natural Resources, March 1996.

Paper Mill Sludge Landfill Cover Construction

Juan D. Quiroz[1], Associate Member, ASCE,
and Thomas F. Zimmie[2], Member, ASCE

Abstract

A considerable amount of experience has been acquired from a number of completed paper mill sludge landfill covers within the United States. This paper focuses on practical guidelines for landfill cover construction using paper mill sludge. The need for these guidelines arises from the unique properties and behavior (high water contents, high organic contents, low shear strengths and a high degree of compressibility) of paper sludges as well as problems that have been encountered during past construction. In general, paper sludge water contents vary from about 100% to 250% and are placed typically 50% to 100% wet of optimum water content while clay water contents are often less than 30% and are placed typically 3% wet of optimum. Hence design issues, construction quality control/construction quality assurance (CQC/CQA) tests, laboratory testing, field compaction, and landfill slope stability have been addressed to facilitate the design and construction process for paper mill sludge landfill covers.

Introduction

The use of paper mill sludge as the hydraulic barrier in landfill cover systems provides an alternative to traditional barrier materials and

[1] Graduate Research Assistant, [2] Professor, Dept. of Civil Engineering, Rensselaer Polytechnic Institute, 110 8th St., Troy, NY 12180-3590, Ph: (518) 276-6941, e-mail: quiroj@rpi.edu and zimmit@rpi.edu.

recycles residual material from the paper making process. Several researchers have explored the ability of paper mill sludge to perform as hydraulic barriers in landfill cover systems (NCASI, 1989, Aloisi and Atkinson 1990, Zimmie et al. 1993, Floess et al. 1995, Moo-Young and Zimmie 1996, Kraus et al. 1997 and NCASI 1997). Paper mill sludges are characterized by high water contents, high organic contents, and low shear strengths and when compared to clay, paper sludge characteristics are in a completely different range. However, in spite of these differences many paper sludge landfill covers have been constructed within the United States and a considerable amount of experience has been acquired. Currently, at least 14 paper sludge landfill covers have been constructed within the United States (NCASI 1997).

This paper focuses on practical guidelines for landfill cover construction using paper mill sludge. The need for these guidelines arises from the unique properties and behavior of paper sludge as well as the problems that have been encountered during past construction which are not common to the construction of typical clay landfill covers. For an overview of typical landfill design and construction see Bagchi (1994). The topics that will be discussed are specific to paper sludge and include an example of typical paper mill sludge characteristics, general design issues and recommendations, preliminary laboratory testing, construction methods and construction quality assurance/construction quality control (CQA/CQC) tests. Construction equipment needs will also be addressed. Construction of paper mill sludge landfill covers differs from that of clay covers, however, complications can be prevented. This is critical to the future of using paper mill sludge as a recycled construction material.

Paper Mill Sludge Geotechnical Properties

Geotechnical characteristics of paper mill sludge are very different from those of typical clays but in general paper sludges do behave similarly (Moo-Young and Zimmie 1996, Kraus et al. 1997 and NCASI 1997). In this paper, the properties of only one paper sludge will be presented as an example. These properties can vary even when dealing with a single paper producing plant, due to paper production changes and changes in wastewater treatment processes.

Zimmie et al. (1993) presented the geotechnical properties of paper sludge produced at the Erving Paper Mill in Erving, MA. The wastewater treatment plant receives about 96% of the influent from a paper mill (a paper recycling plant) and 4% from the town of Erving. The sludge is a blended primary and secondary sludge with an organic content of approximately 50% and a fixed solids content of about 50%. The fixed solids are kaolin clay, used as a filler to produce a smooth finish for paper. The organics, consist largely of tissues and fibers.

Table 1 lists the general geotechnical properties of the Erving Paper Mill sludge. The water contents are the water contents of the paper sludge as produced in the wastewater treatment process. LaPlante (1993) Moo-Young and Zimmie (1996) and Kraus et al. (1997) discuss the difficulties and inconsistencies encountered during Atterberg limits testing for paper sludges. In general, it seems that Atterberg limits are questionable and are not useful indices for paper sludge characterization.

Table 1.   Erving Paper Mill Sludge Properties.*

Property	Value
Water Content (%)	150-250
Organic Content (%)	45-50
Specific Gravity	1.88-1.96
Plastic Limit (%)	94
Liquid Limit (%)	285
Plasticity Index (PI)	191
Compression Index (C_c)	1.24

*After Zimmie et al. (1993).

The compressibility of the sludge is also an important geotechnical parameter in paper sludge landfill cover design. The compression index (C_c) of 1.24 for this paper sludge is high when compared to values for natural clays (typically about 0.20 to 0.40, (Lambe and Whitman 1969)).

Figures 1 through 3 deal with the integrity of paper sludge landfill covers (in terms of hydraulic

conductivity and shear strength). Paper sludge landfill
covers are usually constructed at about 100% wet of
optimum water content where they can be compacted to low
hydraulic conductivities (Moo-Young & Zimmie 1996 and
Kraus et al. 1997). At these water contents and
compaction levels, shear strengths are usually adequate
to maintain stable slopes. However, if increases in
water content or large variations in density occur the
shear strength may decrease and reach critical values.

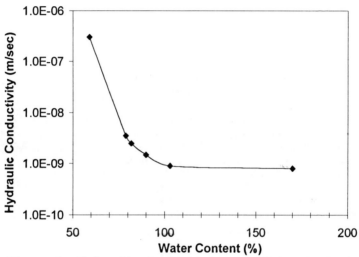

Figure 1. Hydraulic Conductivity vs. Water Content
(Adapted from Zimmie et al. 1993).

The hydraulic conductivity versus water content
relationship for the Erving paper sludge presented in
Figure 1 shows the broad range of water contents for
which a hydraulic conductivity (k) less than 10^{-9} m/s
can be obtained. Notice that this range of water content
is about 50% to 100% wet of the optimum water content
(shown in Figure 2).

Figure 2 shows the standard Proctor curve for the
Erving paper sludge and the zero air voids curve (ZAVC)
(representing 100% saturation). The two curves are
similar. Thus LaPlante (1993) suggested for the water
contents of concern and for placement of the paper
sludge, the ZAVC provides a very close approximation of
the maximum in-place dry density. Proctor tests for

paper sludge generally require several weeks, since they are performed from the wet side (Moo-Young 1995). The paper sludge must be allowed to naturally air dry. This takes time due to the high water contents. Oven drying and rewetting of the paper sludge is unacceptable. This may play an important role in CQA/CQC testing.

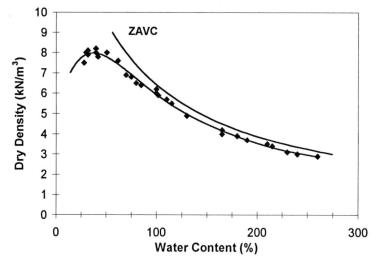

Figure 2. Standard Proctor Curve (Adapted from Zimmie et al. 1993).

Figure 3 illustrates the in-situ undrained shear strength versus water content relationship for a landfill cover using Erving paper sludge. This data was obtained after construction of the paper sludge landfill cover. The peak undrained shear strength was measured using a standard tapered field vane and ASTM D-2573 guidelines. Similar results can be produced from laboratory studies that use triaxial testing or direct shear testing or some other suitable method for determining undrained shear strength. Figure 3 shows the low values of shear strength typical of paper sludge. After examination of the results and field conditions it was determined that at water contents greater than 150% to 160% the shear strength may be approaching critical values.

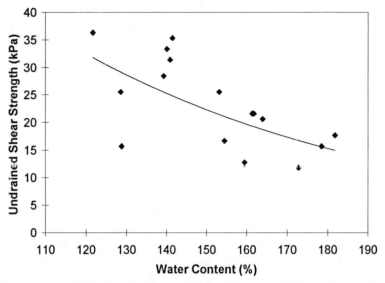

Figure 3. Undrained Shear Strength vs. Water Content,
         for Stable Slopes not Steeper than 3H : 1V.

General Design Considerations

    Identifying the desired dry density and water
content ranges to produce adequate hydraulic
conductivity and shear strength is the first step in
design. Detailed design procedures for paper sludge
landfill covers are beyond the scope of this paper,
however, for design aspects dealing with hydraulic
conductivity of paper sludge barrier see Moo-Young and
Zimmie (1997). For general procedures to account for
factors affecting landfill cover integrity and to
identify dry density and water content ranges for design
see Daniel and Benson (1990). Typically, the main
criteria for directly assessing landfill cover integrity
are dry density, water content and hydraulic
conductivity (undrained shear strength can also be
included). Moo-Young (1995) concluded that the hydraulic
conductivity of paper mill sludges used as landfill
covers is affected by water content, organic content and
consolidation. As a result, other CQA/CQC tests should
also be performed and will be discussed in the
subsequent sections.

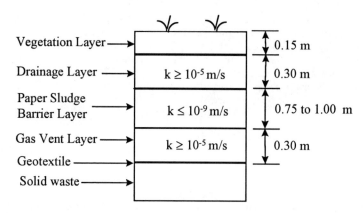

Figure 4. Typical Cross Section.

Figure 4 shows a typical cross section of a paper sludge landfill cover. The typical thickness of the paper sludge barrier layer varies from 0.75 m to 1 m to account for a high degree of expected settlement. The amount of settlement strains experienced in clay covers is low, 2% to 3%, whereas that for paper sludge covers ranges from around 20% to 35% (Zimmie et al. 1993, and NCASI 1997). An important consideration is the installation of a drainage layer above and below the paper sludge to promote consolidation. A geomembrane above the paper sludge barrier layer is not recommended since this will decrease the consolidation rate and prevent migration of any gas generated by the sludge, which if trapped may decrease the shear strength. Also, the interface friction between both materials can reach critical values (create a slick interface) due to the high water contents of paper sludges. Normally a geotextile is placed between the bottom of the impermeable layer and the top of the gas vent layer. This geotextile layer has been omitted in the case of most paper sludge barrier layers, since there is little data available on clogging and filtration of paper sludge-geotextile interfaces. To date no construction problems have been encountered with this interface. Paper sludge can be placed directly on sandy soils with minimal clogging or mixing occurring.

Drainage layers are critical components the cover design since the rate of improvement of paper sludge properties and behavior (i.e., especially hydraulic conductivity and undrained shear strength) depends largely on consolidation. Moo-Young and Zimmie (1996) and NCASI (1997) provide long term data on hydraulic conductivity. Moo-Young and Zimmie (1996) have presented data showing an order of magnitude decrease in hydraulic conductivity to a value less than 10^{-9} m/s of the Hubbardston, MA paper sludge cover over a period of 18 to 24 months.

As noted, an important consideration in paper sludge landfill cover design is slope stability. Due to the relatively low shear strengths of paper sludges, slope stability can become a major issue in overall landfill cover integrity. A factor of safety of 1.5 is reasonable (Sharma and Lewis 1994). Infinite slope stability analyses are suitable since the potential failure mechanism is generally planar. The rule of thumb for paper sludge cover side slopes is a slope of 25%, which is less than the 33% more typical for clay covers. This reduced slope will increase the factor of safety of the slope and is a feasible recommendation for design and construction. The stability of the slope should be verified by suitable slope stability analyses using data similar to that presented in Figure 3. To date no data exists on the use of paper sludge landfill covers in seismic zones, moreover, slope stability will be a governing factor due to the low shear strengths of paper sludges.

In cold regions, a material resistant to freeze-thaw effects may be a primary concern. Moo-Young and Zimmie (1996), Moo-Young and Zimmie (1996b), Moo-Young et al. (1996), Zimmie et al. (1997) and Kraus (1997) present results showing the effects of freeze-thaw cycles on hydraulic conductivity. They all concluded that the effects are similar to those of clay (i.e., an increase in hydraulic conductivity from one to two orders of magnitude). However, sludges are more resistant to freezing than clays due to their high water contents. If freezing and thawing conditions are a concern an additional layer of soil or sludge can be added as a frost protection layer. New York State, for example, requires an additional barrier protection layer that is 0.6 m thick (NYSDEC 1997) to guard against desiccation, frost and root penetration.

Construction Sequence

Construction sequences differing from those required for compacted clays are needed for the placement of paper sludge hydraulic barriers. The construction operations should permit continuous placement of the paper sludge and prevent problems such as slope instabilities that may occur during construction or immediately after construction. The slope instabilities can be caused by large increases in water contents, for example produced by unforeseen heavy rainfalls. The main objective is to construct the paper sludge landfill cover in smaller sections such that large areas of paper sludge are not left exposed. About 0.25 hectare sections is adequate for most landfill closures.

Once the gas vent layer is placed, construction of the paper sludge barrier layer can be initiated. When the full thickness of the paper sludge barrier is constructed, over small sections, then placement of the sand drainage layer should follow as soon as possible. The purpose of immediately placing the sand drainage layer is to confine and protect the paper sludge barrier and to initiate consolidation which helps to stabilize the slopes. The sand drainage layer drains water away from the paper sludge in the case of heavy rains. Heavy precipitation falling directly on exposed sludge can potentially increase the water content, decrease the shear strength, create seepage forces and decrease the factor of safety of the slopes. This method of construction is not typical for clay landfill covers, where often many acres are covered with clay prior to drainage layer placement.

Preliminary Laboratory Testing

Chemical analyses of the paper sludge will probably be the first type of testing performed to evaluate the paper sludge since it determines whether or not the paper sludge is toxic or hazardous. A typical leaching test is the TCLP (Toxicity Characteristic Leaching Procedure) Test.

Preliminary laboratory tests on paper sludges being considered for landfill covers should include Proctor compaction tests, hydraulic conductivity vs. water content tests, undrained shear strength vs. water content (and/or density) tests, consolidation tests, organic contents, water contents and specific gravity tests. As previously noted, Atterberg limit tests are not recommended since the results are questionable. However, most regulatory agencies require Atterberg limits testing, an issue that should be addressed in the process of educating the regulators about paper sludge.

Hydraulic Conductivity Testing

Hydraulic conductivity tests for preliminary or CQA/CQC purposes must be performed since it is the main parameter of concern. These tests can be performed using a triaxial permeability device (Zimmie et al. 1981), utilizing ASTM D-5084. Trimming paper sludge samples causes a high degree of sample disturbance due to the presence of fibers, and should be avoided. Thus, both remolded and undisturbed specimens should utilize molds of the same diameter as that of the triaxial specimen, so that only extrusion is required. In practice, standard Shelby tubes (7.30 cm inside diameter) work well. To obtain undisturbed samples, Moo-Young (1992) suggests that a dynamic sampling process like striking the Shelby tube with a hammer is the best method for high rates of sample recovery and minimal disturbance.

Especially for paper sludge hydraulic conductivity testing, triaxial devices allow for more flexibility and give the tester more options and ultimately more results that can later be used in the paper sludge barrier design process. For example, in an effort to simulate in-situ stresses, landfill covers are often tested using an effective stress of 34.5 kPa. The resulting hydraulic conductivity will be indicative of the hydraulic conductivity achieved immediately after placement. Then, the paper sludge (usually the same specimen) can be retested under an effective stress of 70 kPa for predictive purposes. The hydraulic conductivity obtained under an effective stress of 70 kPa can yield an estimate of the hydraulic conductivity after consolidation (typically a year or so after placement).

The change in hydraulic conductivity due to cover consolidation can be an order of magnitude. Thus, although initial paper sludge hydraulic conductivity may exceed regulatory requirements, it may decrease to an acceptable value after consolidation. This decrease of hydraulic conductivity can be quantified with the results of triaxial permeability tests and consolidation tests. For more information on this topic see Zimmie and Moo-Young (1995).

Field Compaction Methods

There are some advantages to constructing paper sludge covers compared to clay barriers. The range of water contents required to give adequate hydraulic conductivities and shear strengths are much larger than for clays, and the values of water content are much higher. Thus, desiccation is of little concern, a common problem in clay barrier construction. Moreover, paper mill sludge is very manageable and easy to work at these high water contents provided the correct equipment is used. A high compactive effort is not necessary as is the case for clay covers. However, caution should be exercised when placing paper sludge since some special techniques and procedures should be followed.

Clay caps are often constructed using sheep's foot rollers and methods described in NAVFAC DM-7 (U.S. Department of the Navy 1982). However, for paper sludge caps the use of sheep's foot rollers is not recommended since they often become clogged with material and lose their effectiveness. The best method for determining suitable equipment is to prepare trial areas or test plots using different compaction equipment. Experience has shown that paper sludge placement and compaction using low pressure dozers followed by rolling with a smooth small diameter roller pulled by the dozer works best. The roller creates a smooth finish allowing the paper sludge cover to shed water in case of sudden rains. The additional compactive effort produced by heavy rollers or dozers is not as effective as low pressure dozers due to the fact that at high water contents (and very plastic consistencies) additional compactive effort seems to shear the material rather than compact it. Therefore, low pressure dozers and rollers easily compact the sludges since they are nearly saturated and highly plastic.

Moisture conditioning is usually not a problem since the range of working water contents is large as compared to clay. Receiving paper sludge that is too dry should be avoided because adding water to the paper sludge does not create a cohesive material with the same consistency as the original sludge. Material that is too wet can be spread or stockpiled and allowed to air dry until the desired water content is achieved.

Construction Quality Assurance/Quality Control (CQA/CQC)

Two important aspects of landfill cover construction are CQA and CQC. CQC assesses the quality of the material being used for the constructed facility while CQA assesses the quality of the constructed facility. This section will provide recommendations for the type of CQA/CQC tests that should be performed to characterize paper sludge behavior and property variability. The frequency of tests should follow the minimum recommendations outlined by USEPA (1989), the local State regulatory agency or the geotechnical engineer.

Some of the CQA/CQC tests used to monitor paper sludge properties are not standard landfill cover tests, e.g. measuring organic content, specific gravity, shear strength and not using Atterberg limits for characterization purposes. Since the hydraulic conductivity and undrained shear strength are affected by consolidation (i.e., changes in void ratio), the properties that affect void ratio should be monitored closely. Table 2 lists recommended CQA/CQC tests for paper sludge barrier layers.

In-place sand-cone density tests are recommended since the paper sludge densities and water contents are normally out of the range of nuclear density gauges, which are designed for measurements on clays or sands. Also, balloon density tests may yield unreliable results since the paper sludge is very soft and expands with the balloon, resulting in erratic data.

Slope Stability and Shear Strength Control

Special attention should be directed to field CQC/CQA methods which help avoid complications during construction, such as the stability of side slopes. The most critical time is during construction or immediately thereafter, which implies that undrained shear strength is the governing soil parameter.

Table 2. Recommendations for CQA/CQC Testing of Paper Sludge Barrier Layers.

Item	Comment
1. Water Content	60° C oven temperature.
2. Atterberg Limits	Not recommended.
3. Organic Content	CQA/CQC test.
4. Specific Gravity	CQA/CQC test and for void ratio correlations.
5. Dry Density vs. Water Content (Proctor Test)	Compaction test can last several weeks.
6. Hydraulic Conductivity vs. Water Content	Preliminary test for design.
7. Undrained Shear Strength vs. Water Content (and/or Dry Density)	Preliminary test for design-slope stability.
8. Consolidation Test	Settlement calculations.
9. In-Place Sand Cone Density Test	Best method compared to nuclear gauge and balloon method.
10. Laboratory Hydraulic Conductivity for Undisturbed and Remolded Specimens	Test at 34.5 kPa (and 70 kPa for predictive purposes) effective stress. Sample trimming severely disturbs the specimen.
11. Shear Strength	Field vane tests.
12. Chemical Analyses (TCLP Test)	Preliminary and post-closure test.

Experience has shown that the use of a handheld field vane appears to be an excellent approach for acquiring undisturbed peak undrained shear strengths for CQA of the barrier layer. The field vane test is inexpensive since test times are short and simple. This test typically takes about 10 minutes to perform. ASTM D-2573 guidelines and Flaate (1966) discuss factors influencing vane results and should be referenced to

help minimize data variability. A water content sample
(and/or density test) should be taken next to the test
location such that an undrained shear strength vs. water
content (or density) relation can be established and
documented for future reference.

The undrained shear strength can be correlated to
water contents (and even dry density) as presented in
Figure 3. Either through laboratory tests (e.g.,
triaxial compression tests and direct shear tests) or
field vane tests performed on test pads, a minimum
undrained shear strength can be determined and
correlated to water content (and/or density). If a field
vane test measures a low shear strength, lower than the
specified minimum, or if the water content is higher
than specified for adequate strength, then the stability
of the slope may be in question. The ideal situation is
to measure the shear strength and water content as the
barrier is constructed and identify any weak spots. If
any weak spots are identified, then they can be reworked
immediately. This provides additional information on the
density and may even replace the density test after
sufficient results have been obtained to produce a
correlation. Field vane tests certainly show potential
as CQA tests during paper sludge landfill cover
construction. Furthermore, field vane tests can be used
to monitor long term slope stability of paper sludge
landfill covers, as described in the following section.

Long Term Monitoring of Paper Sludge Cover Systems

The use of paper sludge landfill covers is still
generally considered non-conventional, as compared to
the use of compacted clays. Thus, some long term
monitoring will probably be required by regulatory
agencies. Typical post-closure monitoring plans for
paper sludge barrier layers include hydraulic
conductivity, water content, organic content, specific
gravity, consolidation (or settlement data), chemical
analyses and density tests. Post-closure shear strength
monitoring is usually not required. After construction
the paper sludge barrier undergoes a considerable amount
of settlement or consolidation, which generally leads to
a decrease in hydraulic conductivity and an increase in
shear strength. Therefore, it can be said that paper
sludge behavior will improve over time.

In order to track this behavior, the same tests as described for CQC/CQA should be performed until the paper sludge barrier layer has consolidated. Settlement plates installed to measure barrier settlement (or consolidation) are recommended.

Organic contents, to evaluate the organic decomposition of the barrier layer under long term conditions, are of interest. Zimmie and Moo-Young (1995) show that hydraulic conductivity decreases with decreasing organic content. Al-Khafaji and Andersland (1981) performed laboratory vane tests on several paper sludge mixtures and concluded that undrained shear strength decreases with decreasing organic content. However, the results in Figure 3, obtained from a landfill cover constructed over a 4 year period, show that the strength gain due to consolidation far outweighs any strength loss due to organic decomposition. The undrained shear strength values greater than 30 kPa with water contents less than about 140% correspond to the section of the landfill completed in 1993. The rest of the values correspond to more recent cover construction during 1996 and 1997. From these results, the increase in undrained shear strength and decrease in water content is a result of consolidation of the paper sludge barrier. Therefore, it can be concluded that decreasing organic content and consolidation decreases hydraulic conductivity and increases undrained shear strength.

Conclusions

The following conclusions and recommendations can be made for the construction of paper sludge landfill covers:

1. In spite of high water contents, high organic contents, low shear strengths and a high degree of compressibility, paper sludge landfill covers can be constructed.
2. For paper sludge landfill cover design, hydraulic conductivity and slope stability are the primary design parameters.
3. Consolidation of paper sludge decreases hydraulic conductivity and increases shear strength. Thus, the

properties of paper sludge barriers improve with time.

4. The drainage layers of the cover design are critical components since they promote higher consolidation rates.

5. Low pressure construction equipment is required to adequately place and compact the paper sludge barrier.

6. CQA/CQC and post-closure monitoring tests for paper sludge landfill covers (some of which are not common to clay covers) include: water content, density, organic content, specific gravity, consolidation, hydraulic conductivity and shear strength tests. Atterberg limit tests are not recommended.

7. The construction sequence of the paper sludge cover can play an important role in final cover stability during and immediately after construction.

8. Settlement plates in the paper sludge barrier are recommended to track consolidation of the barrier layer.

9. Field vane tests can be used to develop initial slope stability requirements and show potential as a CQA test during placement of the paper sludge barrier.

References

Al-Khafaji, A.W.N. and Andersland, O.B. (1981), "Compressibility and Strength of Decomposing Fibre-Clay Soils," Geotechnique, 31(4), 497-508.

Aloisi, W. and Atkinson, D.S. (1990), "Evaluation of Paper Mill Sludge for Landfill Cover Capping Material," Report Prepared for the Town of Erving, MA by Tighe and Bond Consulting Engineering, Westfield, MA.

Bagchi, A. (1994), Design, Construction, and Monitoring of Landfills. John Wiley & Sons, Inc., New York.

Daniel, D.E. and Benson, C.H. (1990), "Water Content Density Criteria for Compacted Soil Liners," Journal of Geotechnical Engineering, ASCE, 116 (12), 1811-1830.

Flaate, K. (1966), "Factors Influencing the Results of Vane Tests," Canadian Geotechnical Journal, 3(1), 18-31.

Floess, C., Smith, R.J.F. and Hitchcock, R.H. (1995), "Capping with Fiber Clay," Civil Engineering, ASCE, August, 62-63.

Kraus, J.F., Benson, C.H., Maltby, C.V. and Wang, X. (1997), "Laboratory and Field Hydraulic Conductivity of Three Compacted Paper Mill Sludges," Journal of Geotechnical Engineering, ASCE, 123 (7), 654-662.

Lambe, T.W. and Whitman, R.V. (1969), Soil Mechanics. John Wiley and Sons, Inc., New York, NY.

LaPlante, K. (1993), Geotechnical Investigation of Several Paper Mill Sludges for Use in Landfill Covers. Master of Science Thesis. Rensselaer Polytechnic Institute, Troy, NY.

Moo-Young, H.K. (1995), Evaluation of Paper Mill Sludges for Use as Landfill Covers. Ph.D. Thesis. Rensselaer Polytechnic Institute, Troy, NY.

Moo-Young, H.K. and Zimmie, T.F. (1996), "Geotechnical Properties of Paper Mill Sludges for Use in Landfill Covers," Journal of Geotechnical Engineering, ASCE, 122 (9), 768-776.

Moo-Young, H.K. and Zimmie, T.F. (1996b), "Effects of Freezing and Thawing on the Hydraulic Conductivity of Paper Mill Sludges Used as Landfill Covers," Canadian Geotechnical Journal, 33 (5), 783-792.

Moo-Young, H.K., Zimmie, T.F. and Morgan, M.H. III (1996), "Predicting the Level of Frost Penetration into Landfill Covers," Proc., 8th International Conference on Cold Regions Engineering, Fairbanks, AK ASCE, New York, 745-756.

Moo-Young, H.K. and Zimmie, T.F. (1997), "Utilizing a Paper Sludge Barrier Layer in a Municipal Landfill Cover in New York," Testing Soil Mixed with Waste or Recycled Materials, ASTM STP 1275. M.A. Wasemiller and K.B. Hoddinott, Eds., American Society for Testing Materials (ASTM), Conshohocken, PA, 125-140.

National Council of the Paper Industry for Air and Stream Improvement (NCASI) (1989), "Experience with and Laboratory Studies of the Use of Pulp and Paper Mill

Solid Wastes in Landfill Cover Systems," NCASI Tech. Bulletin No. 559, NCASI, New York, NY.

NCASI (1997), "A Field Scale Study of the Use of Paper Industry Sludges in Landfill Cover Systems: Final Report," NCASI Tech. Bulletin No. 750, NCASI, New York, NY.

New York State Dept. of Environmental Conservation (NYSDEC) (1997), 6 NYCRR Part 360 Solid Waste Management Facilities. NYSDEC, Albany, NY.

Sharma, H.D. and Lewis, S.P. (1994), Waste Containment Systems, Waste Stabilization, and Landfills: Design and Evaluation. John Wiley & Sons, Inc., New York.

U.S. Dept. of the Navy (1982), Foundations and Earth Structures, Design Manual 7.2 (NAVFAC DM 7.2), Naval Facilities Engineering Command, Alexandria, VA, 45-50.

U.S. Environmental Protection Agency (USEPA) (1989), "Requirements for Hazardous Waste Landfill Design, Construction, and Closure," EPA/625/4-89/022, USEPA, Cincinnati, OH.

Zimmie, T.F., Doynow, J.S. and Wardell, J.T. (1981), "Permeability Testing of Soils for Hazardous Waste Disposal Sites," Proc., 10th International Conference on Soil Mechanics and Foundation Engineering, Stockholm, Sweden, 403-406.

Zimmie, T.F., and Moo-Young, H.K. (1995), "Hydraulic Conductivity of Paper Sludges Used for Landfill Covers," Geoenvironment 2000; Geotechnical Special Publication No. 46. Y. Acar and D. Daniel, Eds., ASCE, New York, NY, 932-946.

Zimmie, T.F., H. Moo-Young and K. LaPlante (1993), "The Use of Waste Paper Sludge for Landfill Cover Material," Green '93: Waste Disposal by Landfill; Symp. of Geotechnics Related to the Environment. R.W. Sarsby, Ed., Balkema, Rotterdam, 487-495.

Zimmie, T.F., Quiroz, J.D., LaPlante, C.M. and Moo-Young, H.K. (1997), "Comparison of Freeze-Thaw Effects on the Hydraulic Conductivity of Landfill Barrier Materials," Proc., 13th International Conference on Solid Waste Technology and Management, Philadelphia, PA.

Mitigation of Void Development Under Bridge Approach Slabs Using Rubber Tire
Chips

Richard A. Reid[1], M. ASCE, Steven P. Soupir[2], S.M. ASCE,
Vernon R. Schaefer[3], M. ASCE

Abstract

The problem of void development under bridge approach slabs has been
correlated to the use of integral abutment bridges (Schaefer and Koch, 1992). This
void development then causes settlement of the approach slabs leading to the
formation of a bump at the ends of the bridge. The observation of the occurrence of
voids under approach slabs, even in cases where no traffic had yet occurred, led to a
hypothesis of thermally-induced movements of bridge beams/abutment walls as the
mechanism causing the void development. As a result of identification of the
mechanism of void development, changes to the approach system needed to be
made to accommodate this mechanism. One approach to prevent the passive failure
of the backfill soil due to the movement of the bridge abutments is to place a layer
of rubber tire chips between the backfill soil and the back face of the abutments.
This compressible layer can deform with the abutment and prevent passive failure
of the backfill soil.

A scale model study has been performed using shredded tire chips in the
backfill system. This paper will present the backfill design, instrumentation, testing
methodology and data analysis of this model test. The data analysis includes a
comparison of data obtained from this model test with previous tests conducted of a
backfill system without the incorporation of rubber tire chips.

[1]Assistant Professor, Civil and Environmental Eng. Dept., South Dakota State
University, CEH 106, Box 2219, Brookings, SD 57007-0495, 605-688-6368,
Reidr@mg.sdstate.edu

[2]Graduate Research Assistant, Civil and Environmental Eng. Dept., South Dakota
State University, CEH 110, Box 2219, Brookings, SD 57007-0495, 605-688-6369,
R6ap@sdsumus.sdstate.edu

[3]Professor, Civil and Environmental Eng. Dept., South Dakota State University,
CEH 110, Box 2219, Brookings, SD 57007-0495, 605-688-6307,
Schaefev@ur.sdstate.edu

Introduction

The problem of void development under bridge approach slabs has been correlated to the use of integral abutment bridges (Schaefer and Koch, 1992). The observation of the occurrences of voids under the approach slabs, even in cases where no traffic had yet occurred, led to a hypothesis of thermally induced movements of the bridge abutment as the mechanism causing the void development.

To investigate methods to reduce pressures on the bridge end backfill, the South Dakota Department of Transportation (SDDOT) sponsored a study of alternative backfill designs to be evaluated through model studies. A model test using a vertical layer of rubber tire chips has been performed using similar testing techniques and procedures as in a previous study by Schaefer and Koch (1992), which incorporated the use of a backfill system containing entirely select granular backfill. Using similar testing methods for current testing allowed for a comparative analysis of test results for the two types of backfill designs.

Previous Research

The use of integral abutment bridges has grown steadily over the past 35 years to eliminate the continuous maintenance problems associated with bridge joints. The thermal movement of integral abutment bridges was the focus of a 1973 study conducted at South Dakota State University (SDSU) for the SDDOT (Lee and Sarsam, 1973). A model bridge, with an integral abutment supported on piles, was constructed and tested to monitor the thermally induced stresses on the bridge during and after construction. The model was loaded by applying hydraulic jacks to simulate temperature expansion and contraction of the bridge. Cycles were run to simulate construction stages and different seasons of the year. Results showed that integral abutment systems provided an alternative to relieve costly maintenance problems associated with bridge joints and in bridge structures.

Several years after implementation of integral abutment designs in South Dakota, cracks began to develop in the bridge approach slabs. Preliminary investigations determined these cracks were caused by the development of voids under the approach slab (Schaefer and Koch, 1992).

A field study of over 130 bridges in South Dakota showed that most integral abutment systems suffered void development problems. Of the 79 post-1980 integral abutment bridges observed, voids under the approach slab ranging from less than 2.5 to 35.5 cm in height were observed in 57 cases. An additional 18 of the 79 integral abutment bridges had previously been mudjacked from what was observed to be the development of a void under the approach slab. Only one of the 79 integral abutment bridges was conclusively determined not to have developed a void (Schaefer and Koch, 1992).

Some of the problems that arise from the development of voids under the approach slab are approach slab cracking and embankment cracking. Cracking of the approach slab occurs due to unsupported loading of the slab from the development of voids and was evident in approximately one-half of the bridges inspected. In most instances, the cracking occurred at the end of the dowel bars that protruded from the bridge abutment section into the approach slab section. It was determined that a stress concentration develops at the ends of the dowel bars, therefore causing cracks to develop in the concrete approach slab. Cracking of the approach embankment typically developed from the contact of the wing wall and the abutment while radiating out at 45 degrees. From the field study, it could not conclusively be determined that integral abutments where the cause of cracking in the embankment material. However, observations showing no such embankment cracking in non-integral abutments pointed to abutment movement as a highly likely source.

In order to confirm that the primary mechanism causing the development of voids under the approach slab was thermally induced movements of the integral abutment, a model study was performed in order to replicate abutment movement. It was determined that the primary mechanism causing void development under the approach slabs is the thermal-induced movements of the integral abutments. This mechanism is also responsible for the development of cracks in berms and approach embankments as the abutment expands and contracts. Void development effects are compounded by the combination of several mechanisms such as embankment bulging as the backfill deforms, approach slab uplift, backfill densification as particle breakage occurs, and backfill deformation as passive failure occurs in the backfill. Cracking of the approach slab is a direct consequence of the loss of support under the approach slab, which affects the flexural stresses induced in the slab.

Model Study

In order to reduce construction costs and eliminate the mechanisms causing voids, alternative solutions were sought. Model tests in which alternative backfill materials, such as lightweight fill materials, are being investigated to determine possible backfill applications in reducing void development under the approach slab. The model used is the same structure used in the 1973 study to determine the stresses developed in integral abutment bridges. It is composed of a partial bridge section, jacking abutment, and integral abutment approximately 5.8 m wide and 9.14 m in length. The abutment is oriented east-west and is scaled down to approximately one-half width, but is of full depth and pile supported. The abutment has a 3.6-m approach sill and a 1-m wing walls. The approach system was designed to reflect SDDOT practices at that time. It contained a backfill material consisting entirely of a select granular material. This expensive backfill was used since it had been shown to reduce void development due to its greater passive resistance. The select granular backfill contained material with less than five percent passing the #200 sieve and extended the depth of the abutment and typically back at a 1:2 slope. The approach slab was composed of four individual sections for ease of handling. Each section was 1.8-m wide, 3-m long, and 15.2 cm thick. The total approach slab dimensions are a 3.6-m width, 6-m length, and 15.2-cm thickness. Shoulders were two feet wide with 2H:1V side slopes and the fill embankment was designed to be constructed to a 6H:1V slope in accordance with standard SDDOT practice.

Backfill Design

Research prior to the implementation of the model test had revealed that the use of lightweight fill materials in approach embankments had not been reported. A review of engineering characteristics such as compressibility, shear strength, compaction behavior, resilient modulus, hydraulic conductivity, and environmental impact identified shredded rubber tire chips as a likely candidate for use in an integral abutment backfill design application. The use of rubber tire chips as a fill material has previously been investigated by Ahmed and Lovell (1993), Humphrey, et. al, (1993), Upton and Machan (1993), Edil and Bosscher (1992), Manion and Humphrey (1992), Engstrom and Lamb (1994), and Ahmed (1993). An evaluation of these materials has shown that research in using rubber tire chips to date has primarily been in the area of lightweight fill on weak soils with little emphasis on the use of rubber tire chips in a vertical layer to absorb differential movement of the integral abutment. There has been one application of rubber tire chips as a compressible inclusion on a major earth retaining structure in Maine (Whetten, et. al, 1997). It was reported that a 1-m-wide zone of rubber tire chips against the abutment wall of a rigid frame bridge was used to lower the compression of the surrounding soil. The purpose of this application was to reduce the magnitude of earth pressures applied to the walls. Due to the elastic capabilities of rubber tire chips, a design scheme using a vertical layer of rubber tire chips between the integral abutment and the granular backfill material was adopted for use in the model test. Compression tests on rubber tire chips had shown that tire chips are highly compressible upon the first loading cycle but decreases during subsequent loading. It was anticipated that these elastic properties of rubber tire chips would allow for the lateral movements of the bridge abutments without affecting the select granular backfill. The rubber tire chips would essentially act as a cushion for the integral abutment to move into and away from the select fill with temperature induced movements, therefore not allowing the select granular backfill to be affected by induced movements of the abutment wall.

The benefits of using rubber tire chips as a backfill material include reducing the large number of rubber tire stockpiles found in South Dakota, reduced weight of fill, reduced settlements, preventing development of pore pressures during loading of fills, conservation of energy and natural resources, and allowing the use of a substitute material for the more expensive conventional fill materials.

The SDDOT provided the shredded rubber tire chip material used in the backfill design. Shredded tire chip sizes varied considerably from one to six inches and protruding wires were evident in each chip.

Model Construction

To perform the model test with the use of a vertical layer of rubber tire chips behind the abutment wall, the existing backfill material had to first be excavated.

Figure 1 illustrates the excavated portion of the backfill system. The four sections of the approach slab were separated and removed with a front-end loader. The select granular backfill was then removed to a depth of nearly 1.7-m down to the underlying drainage fabric away from the abutment wall to 0.6 m behind the inclinometer casings using a front-end loader. From this point, the backfill was removed at approximately a 1H:1V slope up to the top of the existing embankment. The embankment material was kept separated from the select granular backfill as much as possible.

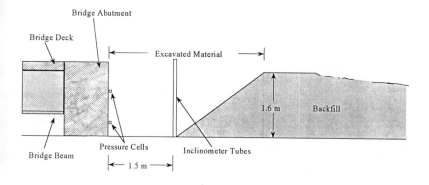

Figure 1. Illustration of excavated portion of model bridge backfill system.

When the excavation was completed, construction of the backfill system began by placing thick cardboard approximately 40 cm from the abutment wall. Due to the small confining space of the rubber tire chip layer, an initial rubber tire chip layer of 40 cm was chosen in order to compact the rubber tire chips by extensive compaction of backfill material against the cardboard material. Thirty centimeter layers of select granular backfill and embankment material were then placed and compacted to greater than 95% compaction, conforming to SDDOT specifications, which was performed with a jumper compactor. Observations of the compacted material were made in areas adjacent to the cardboard separator. It was observed in the first couple of layers that the backfill was indeed compacting the rubber tire chip layer, compressing the layer to about 30.5 cm. This method of compaction appeared to produce sufficient compaction in the rubber tire chips. Research by Ahmed (1993) found that the optimum density could be achieved in a pure rubber tire chip configuration, especially when large amounts of protruding wires are present, with a modest amount of compactive effort. Therefore, it was deemed acceptable to continue construction in this manner. Layers of soil were placed and compacted in this manner until the backfill system was brought up to grade. The approach slab sections were then placed, connected together, and then connected to the bridge deck with 2-cm bolts. Elevation points were established around and on the approach slab to monitor vertical movement of the embankment material and approach slab during testing. An illustration of the constructed model bridge backfill system is shown in Figure 2.

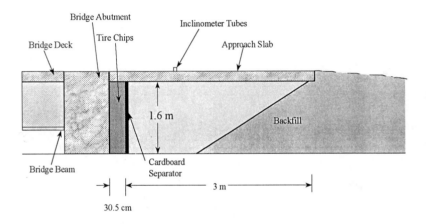

Figure 2. Constructed model bridge backfill system using a 30.5-cm layer of rubber tire chips.

Testing Methodology

Movement of the backfill system, embankment material, and approach slab were monitored in the same manner as with the previous model study performed by Schaefer and Koch (1992). An illustration of the instrumentation plan is shown in Figure 3.

Loading and testing procedures for abutment movement cycles were also followed as in previous research. A review of temperature variations around the state was made to determine the amount of abutment movement that should be induced to the backfill system. Data from around the state showed that the maximum variations in temperature experienced at different locations were very similar and therefore, the maximum temperature change data from Brookings, SD was used in computing the cyclic movements of the integral abutment.

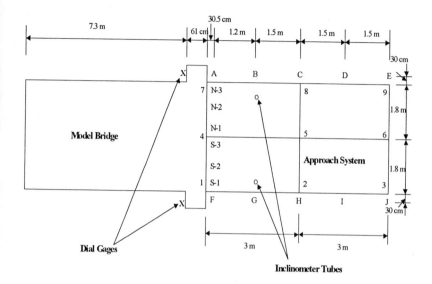

Figure 3. Plan view of approach system showing locations of instrumentation and measuring points. (A, B, C, D, etc. represent embankment stake locations. S-1, N-1, etc. represent void measurement hole locations. 1, 2, 3, etc. represent elevation points on the approach slab.)

This data was then used to calculate movements corresponding to temperature changes with respect to normal, high, and low temperatures. A 91-m bridge was assumed as an average length and the abutment movements were calculated for one-half of this length. An initial movement cycle of May normal temperature was assigned as zero and a pattern of movement was determined. Since it was desired to perform the model test for a one-year cycle of abutment movement, a one-day cycle of movement was developed for each month. Each cycle represented abutment movement from the normal, to the high, to the low of each month, and then to the normal of the next month. A one-year pattern of movement used for testing can be seen in Figure 4.

Loading was performed with two 1800-KN hydraulic jacks for expansion cycles and one 535-KN hydraulic jack was used for contraction cycles. Testing of the backfill system involved moving the abutment with the hydraulic jack equipment until the desired range of movement was accomplished for each cycle. Readings of the inclinometers, pressure cells, embankment stakes, void development holes, and approach slab were taken at each step of the cycle. Testing took approximately one month to complete.

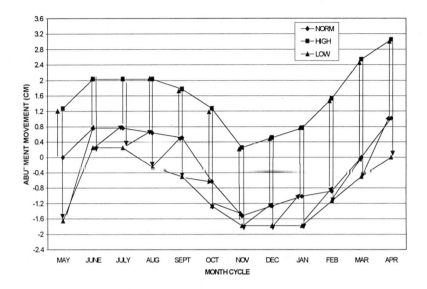

Figure 4. Monthly abutment movement cycles developed from temperature data for May 1990 through April 1991 to simulate a one-year cycle of abutment movement.

Past testing methods were adopted in order to allow for a comparison of results using the vertical layer of rubber tire chips with a backfill composed entirely of a select granular backfill. Selected test results are shown to illustrate essential points. These results will be compared with past research whenever possible to provide a more complete analysis of model test results.

Void Development

The primary reason for conducting the model study was to determine if thermal expansion and contraction has a direct effect on the development of voids under the approach slab. Six holes were bored through the approach slab along a line 30.5-cm from the abutment in order to monitor the void space development underneath the approach slab during testing cycles.

Data from the model test using a vertical layer of rubber tire chips between the abutment and select granular backfill showed that a void became evident and increased or decreased as the abutment contracted and expanded while growing progressively larger with each cycle. Data from holes in the center and at the edges of the approach slab revealed that void development near the outside of the approach slab was larger than near the center with a final void development of approximately 2.8 cm on the outside. An illustration of data collected on the outside of the approach slab for the measurement of void development under the approach slab in the case for the vertical layer of rubber tire chips model test is located in Figure 5.

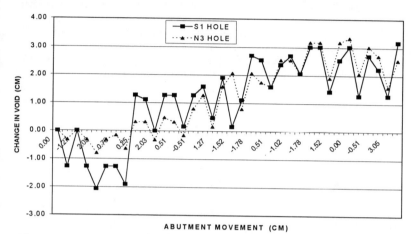

Figure 5. Void development under the approach slab in the case incorporating the use of a vertical layer of rubber tire chips for holes S1 and N3.

It has been theorized that the contributing factors which allow for the development a void under the approach slab in the case where a rubber tire chip layer was used arise from the settlement of the rubber tire chips as the abutment was moved cyclically. Settlement occurs in areas of the backfill adjacent to the rubber tire chip layer, which produced a void under the approach slab. It is anticipated that while increasing compaction of the rubber tire chips, using a different type of rubber tire chip, or even a smaller layer may reduce the amount of void development, some degree of void development would most likely occur.

Data from the model test not using a vertical layer of rubber tire chips showed very similar results but with slightly larger values for the total development of voids. This test revealed an increase in final void development of approximately 3.3 cm.

Earth Pressure Development

Earth pressures imposed on the backfill were measured by two total pressure cells. The pressure cells were mounted in the center of the abutment wall at depths of two and four feet from the top of the abutment. Two Slope Indicator model 51482 total pressure cells with model 514178 transducers were placed on the abutment wall. The cells are flat pressure cells 23 cm in diameter and 1 cm thick. The pressure cells are constructed of stainless steel and are filled with fluid. The fluid allows for the soil pressure on the flat walls of the cell to be converted to fluid pressure and measured by a pneumatic pressure piezometer. The standard pressure range of these cells is 7 to 2068 kPa. The cell pressures were read with a Slope Indicator Model 211 pneumatic indicator with a 0.15 percent digital readout gauge.

In the model test using a vertical layer of rubber tire chips, cyclic movement of the abutment revealed a relationship between abutment movement and changes in earth pressure, but at a low level. In general, pressures developed in the backfill corresponded to abutment movement. The larger the movement for a cycle, the larger the pressures that developed. The highest pressure of approximately 51.7 kPa occurred on the bottom pressure cell in the April high cycle with abutment movement near 3 cm from initial. A pressure decrease was not evident during periods of delay after the backfill was loaded, which had occurred in the previous model test. This is most likely due to the low stresses applied to the backfill system from the incorporation of the rubber tire chip layer. In general, the backfill was not substantially loaded, for the rubber tire chip layer absorbed the abutment wall movement.

In the case where the entire backfill was composed of a select granular backfill, data again clearly showed a relationship between the abutment movements and changes in earth pressure, but on a much larger scale. The highest pressure of approximately 386 kPa occurred on the bottom pressure cell in the March high cycle with abutment movement near 2.5 cm from initial. Pressure cell response also showed that earth pressure redistribution occurred during abutment movement steps. A pressure decrease was evident during periods of delay after loading. This is due to a redistribution of stresses occurring in the backfill.

Inclinometer Movements

Lateral and longitudinal movements of the backfill were monitored with an inclinometer. The inclinometers were positioned approximately five feet away from the abutment and in line with the integral abutment piles in order to possibly record pile movement effects on the backfill. Inclination readings were taken at frequent intervals of depth and were then converted to displacements. The inclinometers in the backfill were oriented to record the lateral and longitudinal movement of the backfill with respect to the initial readings.

Where a vertical layer of rubber tire chips was used in the backfill system, very little movement in the lateral and longitudinal directions occurred. The largest movements recorded with the inclinometer tubes in the backfill in either direction were less than 2.5 mm. In general, the backfill material deflected in a small amount with subsequent cyclic abutment movement, however very little movement occurred. Deflection occurred more near the bottom, which indicates that the abutment supporting piles affected the subsoil underneath the backfill material more than the abutment wall moving into the backfill and embankment materials.

In the case of the backfill system composed entirely of select granular material, little movement occurred in the lateral direction while more substantial movements were evident in the longitudinal directions. The north inclinometer recorded a final lateral expansion of only 2-mm while the south inclinometer recorded slightly more with a final lateral expansion of 4.8 mm. Although these values are relatively small, there is evidence to support that the cycling of the abutment wall is expanding the embankment outward.

In the cases for the longitudinal deflections, the north and south inclinometer had a final deflection of approximately 5.1 cm away from the abutment while the maximum deflections occurred after the April high abutment movement of 3 cm and deflection was measured to be approximately 6.6 cm away from the abutment. Figure 6 illustrate the inclinometer data collected for each model test. Data shows that longitudinal movements of the backfill cycled as the abutment expanded and contracted. It was evident that while mimicking large springtime temperature changes, significant additional cycle and permanent deformations occurred in the backfill and the embankment soils.

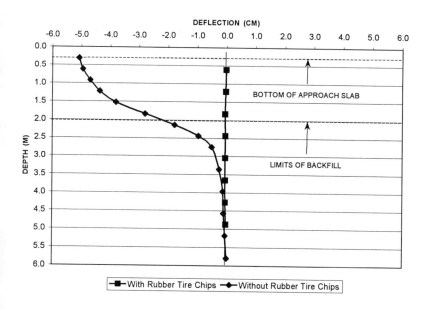

Figure 6. Longitudinal movement of south inclinometer for April low cycle with and without a layer of rubber tire chips.

Approach Slab Movements

Triangulation with theodolite measurement plates was used to monitor the bridge abutment movement during loading. This allowed for the movements of the abutment to be monitored to ensure that equal movement occurred on each side and followed cyclic movements that had previously been determined.

For the model test incorporating the use of a vertical layer of rubber tire chips, the elevation changes in the approach slab were not apparent. This seems to correlate with deflection data in the backfill where little movement in the backfill material was occurring.

In the case of a backfill containing entirely select granular backfill, trends developed in the data where approach slab center stations showed some elevation change with increases of 0.6 to 1.3 cm. This was attributed to the large abutment expansions where upward movement of the backfill occurs during expansion cycles. An apparent final change in elevation of about 0.6-cm occurred for the middle stations on the approach slab. This was calculated to represent approximately seven cubic feet of volume change in the backfill.

Embankment Stake Movements

Stakes were driven into the embankment fill along the approach slab and monitoring locations were set on the approach slab. This allowed for the movements of the embankment and approach slab to be monitored with a level survey and to determine approximate lateral and longitudinal movements of the each stake. Movement of the stakes was determined by using a tape measure. Measurements were made of the longitudinal and lateral movements of the stakes as the abutment was expanded and contracted.

The current model test using a vertical layer of rubber tire chips showed negligible movement of the embankment stakes occurred, with most of the movement occurring in the stakes closest to the abutment. Again, this data shows that the layer of rubber tire chips absorbs a large portion of the induced movements from the integral abutment. Lateral expansion of the embankment material was not evident in this test.

The model test without a rubber tire chip layer showed that the stakes closest to the abutment displayed the most movement in both directions and moved cyclically with abutment movement. All movements were within 1.3-cm showing a small of lateral expansion. In general, measured stake movements on the embankment soils expanded laterally during the test in relatively small amounts.

Conclusions

The occurrence of voids under the approach slab in bridges in South Dakota is very high. Field studies conducted showed that the voids varied from less than 2.5 cm to as large as 35.6 cm. The field studies also revealed a high correlation between void development and integral abutment bridges. A model study was conducted to determine the mechanisms responsible for the development of voids under the approach slab. Based on results of past field test studies and a model study, the mechanism of void development under the approach slabs is due to thermal-induced movements of the integral abutments. This was also identified as the mechanism responsible for cracks in approach embankments.

Incorporating of a vertical layer of rubber tire chips has shown that such a layer between the abutment and the backfill system will inherently prevent longitudinal and lateral movements in the backfill while relieving relatively high earth pressures. Although it has become evident that settlement of the rubber tire chip layer due to the cyclic movements of the integral abutment produce settlement in soils adjacent to the rubber tire chip layer. This settlement allows a void to develop under the approach slab in a similar degree as in the case where no rubber tire chip layer had been used. The lack of compaction of the vertical layer of rubber tire chips between the integral abutment and the bridge end backfill may have lead to void development under the approach slab. The tire chip layer greatly relieves passive earth pressures on the retained fill, which reduces the mechanisms causing void development.

Acknowledgments

This research was funded by the South Dakota Department of Transportation, project SD96-02. The support of the SDDOT is sincerely appreciated.

The contents of this report reflect the views of the authors who are responsible for the facts and accuracy of the data presented herein. The contents do not necessarily reflect the official views of policies of the South Dakota Department of Transportation, the State Transportation Commission, or the Federal Highway Administration. This report does not constitute a standard, specification, or regulation.

References

Ahmed, I., "Laboratory Study on Properties of Rubber Soils," Joint Highway Research Project Engineering Experiment Station, Purdue University, Indiana Department of Transportation, U.S. Department of Transportation Federal Highway Administration, Project No. C-36-50L, May 1993.

Ahmed, I., and Lovell, C.W., "Rubber Soils as Lightweight Geomaterials," Transportation Research Record 1422, 1993, pp. 61-70.

Ahmed, I., and Lovell, C.W., "Use of Waste Products in Highway Construction," Environmental Geotechnology, M.A. Usmen and Y. B. Acar (Editors), A. A. Balakema/ Rotterdam, Brookfield, 1992, pp. 409-418.

Edil, T.B., and Bosscher, P.J., "Development of Engineering Criteria For Shredded Waste Tires In Highway Applications," Division of Highways, Wisconsin Department of Transportation, Final Report, Project No.: WI 14-92, September 1992.

Engstrom, G.M., and Lamb, R., "Using Shredded Waste Tires as a Lightweight Fill Material for Road Subgrades," Minnesota Department of Transportation Materials Research and Engineering, Project No.: 9PR6006, April 1994.

Humphrey, D.N., Sandford, T.C., Cribbs, M.M., and Manion, W.P., "Shear Strength and Compressibility of Tire Chips for Use as Retaining Wall Backfill," Transportation Research Record 1422, 1993, pp. 29-35.

Lee, H.W. and Sarsam, M.B., "Analysis of Integral Abutment Bridges," South Dakota Department of Transportation Study No. 614 (70), Final Report, March 1973.

Manion, W.P., and Humphrey, D.N., "Use of Tire Chips as Lightweight and Conventional Embankment Fill," Technical Services Division, Maine Department of Transportation, Technical Paper 91-1, May 1992.

Schaefer, V.R., and Koch, J.C., "Void Development Under Bridge Approaches," South Dakota Department of Transportation Office of Research, Final Report No.: SD90-03, November 1992.

Slope Indicator Co., Instruction Manual Digitilt Inclinometer-5030090 Instrument, Slope Indicator Company, Seattle, Washington, 1984.

Slope Indicator Co., Instruction Manual Model 211 Pneumatic Pressure Indicator 51421199 Instrument, Slope Indicator Company, Seattle, Washington, 1995.

Upton, R.J., and Machan, G., "Use of Shredded Tires for Lightweight Fill," Transportation Research Record 1422, 1993, pp. 36-45.

Whetten et. al., "Rubber Meets The Road In Maine". Civil Engineering, September 1997, pp. 60-63.

Tire Shreds as Lightweight Fill for Embankments and Retaining Walls

Dana N. Humphrey[1], Nathan Whetten[2], James Weaver[3], Kenneth Recker[3], and Tricia A. Cosgrove[4]

ABSTRACT

Use of tire shreds in three highway projects is described. In the first project, tire shreds were used as a compressible inclusion to reduce pressures on a rigid frame bridge. Earth pressures were reduced by more than 50%. In the second project, a 4.3-m thick zone of tire shreds was used as lightweight fill to improve global stability of a bridge approach fill founded on weak clay. In addition, the tire shreds reduced horizontal pressure on the bridge abutment. In the third project, two layers of tire shreds, each up to 3.05 m thick, were used as lightweight fill for a highway embankment founded on weak clay. These were the first projects to incorporate design features to minimize internal self-heating of tire shred fills. Measured temperatures showed that no deleterious self-heating occurred. These projects demonstrate that tire shreds can be used as lightweight fill for retaining walls and embankments. Moreover, properly designed tire shred fills do not experience a deleterious self-heating reaction.

INTRODUCTION

Tire shreds are scrap tires that have been cut into 30 mm to 300 mm pieces. They have a compacted unit weight that is one-third to one-half that of conventional earth fill. Moreover, tire shreds are durable, have a negligible impact on groundwater (Humphrey, et al., 1997b) and are low-cost making them an attractive lightweight fill for embankments constructed on weak soils. Tire shreds also produce low earth pressure on retaining walls (Tweedie, et al., 1998a,b).

This paper describes three projects constructed in Maine that used tire shreds as lightweight embankment fill and retaining wall backfill. In the first project, tire shreds were used as a compressible inclusion to reduce earth pressures on a rigid

---

[1] Professor, Dept. of Civil and Environmental Engineering, University of Maine, 5711 Boardman Hall, Orono, ME 04469-5711, 207-581-2176, dana.humphrey@umit.maine.edu
[2] Senior Geotechnical Engineer, Haley & Aldrich of New York, 189 North Water Street, Rochester, NY 14604-1151
[3] Vice President, Haley & Aldrich, Inc., 500 SouthBorough Dr., South Portland, ME 04106-6903
[4] Geotechnical Engineer, Bowser-Morner, PO Box 838, Toledo, OH 43697-0838

frame bridge. In the second project, tire shreds were used as lightweight fill behind a bridge abutment to improve the safety factor for global slope stability and reduce lateral pressure on the abutment. These projects are part of a new bypass connecting U.S. Route 1 in Brunswick, Maine to Interstate 95 in Topsham, Maine. In the final project, tire shreds were used as lightweight fill for a bridge approach embankment founded on a weak clay foundation. This project is located in Portland, Maine, and is part of a new interchange being constructed by the Maine Turnpike Authority. These projects show how the properties of tire shreds can be used to overcome difficult design problems and reduce overall construction costs. Some 1.7 million scrap tires were put to a beneficial use on these projects.

RIGID FRAME BRIDGE

The rigid frame bridge is part of the new Topsham-Brunswick Bypass Project, located in Topsham, Maine. It was designed by T.Y. Lin International of Falmouth, Maine, the lead consultant to the Maine Department of Transportation (MDOT) for the project. Geotechnical services were provided by Haley & Aldrich of South Portland, Maine. The bridge provides an overpass for the new Route 196 bypass over a single track railroad owned by MDOT. The bridge geometry is partly defined by the bypass alignment which intersects the railroad at an acute 24 degree angle. The bridge structure spans the railroad providing a 7.62 m wide by 7.01 m high clear opening for railroad traffic. Approach fill heights are up to 10.98 m and the length of the rigid frame tunnel is 93 m. The aspect ratio of the bridge allowed the rigid frame to be designed as a "long barrel" structure. Construction of the bridge began in 1994 and the bridge was open to traffic in November, 1996.

Bridge Foundation Design

Subsurface conditions consisted of 11.3 to 19.2 m of loose to medium dense marine sand overlying 8.8 to 13.4 m of soft to medium stiff marine silty clay, then 0.8 to 3.0 m of glacial till followed by bedrock. Post-construction settlement ranging from 125 to 300 mm was expected due to settlement of the marine clay in response to approach embankment loading. Supporting the bridge abutments on shallow foundations above the clay would have imposed additional stresses on the clay, resulting in added settlement. Therefore, the bridge was supported on piles driven to penetration resistance in glacial till or on bedrock. The abutments were supported on two rows of steel H-piles. Battered piles (1H:4V) spaced 1.83 m on center were driven along the front row for lateral load resistance. Vertical piles spaced 3.66 m on center were driven along the back row as shown in Figure 1.

Lateral Earth Pressures

Rigid frame structures are generally designed for "at-rest" (K_o) conditions. An equivalent fluid unit weight of soil of 0.96 Mg/m^3 was recommended, assuming an at-rest coefficient of earth pressure of 0.5 and no water pressure. As the design progressed, it was determined that a reduction in equivalent fluid pressure would

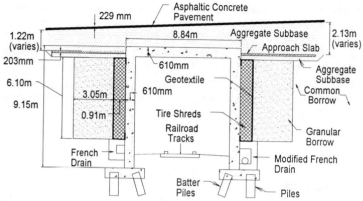

Figure 1. Cross section of rigid frame bridge showing location of tire shred zone.

benefit the project and allow it to be built at a lower cost. An equivalent fluid unit weight of 0.58 Mg/m^3, typical of an "active" earth pressure condition, was targeted for design.

One option for reducing lateral loads on the bridge was lightweight backfill such as expanded shale aggregate or tire shreds. However, this option was costly since it would have required lightweight fill over the full height of the wall and within a 45-degree plane sloped outward and upward from the bottom of the wall. An alternative solution was to place a 0.91-m wide zone of compacted tire shreds as a "compressible inclusion" directly against the abutment walls, between the walls and the adjacent approach fill. The purpose of this limited thickness of tire shreds was to compress horizontally as the approach fill was placed. This would allow the adjacent soil fill to strain sufficiently to mobilize its strength, changing the lateral earth pressure from at-rest to active conditions, thereby reducing the horizontal stress applied to the bridge abutments.

Compressible Tire Shred Inclusion

A 0.91-m wide zone of tire shreds was placed against each abutment wall as shown on Figure 1. The bottom of the tire shred zone was near the top of a French drain. It would have been desirable to start the tire shred zone at the base of the wall, however this was not possible as the French drain had already been constructed as part of an earlier contract. The top of the tire shred zone extended to 305 mm below the bottom of a concrete approach slab. The tire shreds were separated from the adjacent granular borrow backfill by woven geotextile. This prevented the soil from being washed into the tire shreds which could lead to long term settlement. The adjacent zone of granular borrow was 3.05 m wide. Common borrow (marine sand) was used as backfill beyond the limits of the granular borrow.

The tire shreds were made from a mixture of steel and glass belted tires. They were required to have 100% pass the 203-mm square mesh sieve, a minimum of 95% (by weight) passing the 76-mm square mesh sieve, a minimum of 50% passing (by weight) the 51-mm square mesh sieve, and a maximum of 20% passing (by weight) the 4.75-mm (No. 4) sieve. These were termed Type A shreds.

The first step in constructing the tire shred zones was to begin placing the geotextile. The seams between adjacent rolls were specified to be vertical with a minimum 457-mm overlap. No horizontal seams were allowed. Thus, the contractor cut lengths of geotextile that were long enough to reach from the bottom to the top of the tire shred zone. The geotextile was laid out at the bottom of the tire shred zone and lapped 457-mm up the face of the concrete wall. The first lift of tire shreds was placed on the geotextile to hold it in place. Excess geotextile was rolled up at the interface between the tire shreds and granular borrow. The geotextile was unrolled as the fill elevation was raised. The geotextile was placed on the east side first with the specified 457-mm overlap between adjacent vertical strips of geotextile. After placing only a few lifts of tire shreds, it became apparent that wrinkles in the geotextile combined with undulations in the contact between the tire shreds and granular borrow reduced the overlap to significantly less than the specified value. The contractor had to sew on additional widths of geotextile to maintain the overlap. For the tire shred zone on the west side, the contractor started with a 914-mm overlap. This allowed sufficient leeway to maintain the specified 457-mm overlap throughout the full height of the tire shred zone. At the top of the tire shred zone on both sides, the geotextile was lapped 457-mm against the face of the concrete wall.

Tire shreds and adjacent granular borrow were raised in 203-mm lifts. Generally, 203-mm of granular borrow was placed first and then compacted by a smooth-drum vibratory roller. Then, a lift of tire shreds was placed using a front end loader and spread by two laborers using rakes. Each lift of tire shreds was compacted by four passes of a walk-behind vibratory tamping foot roller with an operating weight of 1250 kg. Previous experience has shown that vibratory plate compactors are ineffective for compacting tire shreds (Tweedie, et al., 1998a,b). This process was continued until the full height of the tire shred zones was placed. The top of the completed geotextile wrapped tire shred zone was covered with 305 mm of granular subbase aggregate. This formed the base for the cast-in-place approach slab. After casting the approach slab, 1.22 to 2.13 m of subbase aggregate was used to bring the road up to the grade required for paving.

Instrumentation

Pressure cells, soil strain meters, slope indicators, and temperature sensors were installed to monitor lateral earth pressures on the wall, as well as the temperature and movement within the tire shred zone. Eight pressure cells were used to measure the horizontal stress generated by the backfill. They were installed at stations 306+25 and 306+40 on the east abutment wall with four cells at each station. Soil was placed against the lowest cell at each station. Tire shreds were placed against the remaining cells. Two types of pressure cells were used: Rocktest model EPC and Rocktest model TPC. Compared to the model EPC, the TPC is stiffer and the readings are less sensitive to temperature changes. The authors were unsure as to which would give the most reliable readings, so four of each type were used. Information on calibration of the pressure cells may be found in Humphrey, et al. (1997a), and Cosgrove and Humphrey (1998).

Horizontal compression of the tire shred zone was measured using eight Rocktest model COR-P soil strain meters (convergence meters). They were placed perpendicular to the wall with one end bearing against the wall and the other end extending about 762 mm out into the tire shred fill. One meter was placed at the same elevation as each pressure cell. Two slope indicator casings were placed at station 306+10. They were offset about 1.1 m and 3.5 m from the interface between the tire shreds and the granular borrow. They were used to monitor horizontal movement of the soil adjacent to the tire shred layer.

Thermistors (Rocktest Model TH-1) were placed in the backfill adjacent to each pressure cell. In plan, they were located about 457 mm from the face of the walls. Two are located in granular backfill below the tire shred layer and six are located in the tire shreds. These were used to monitor the temperature of the tire shreds and granular borrow backfill.

Measured Horizontal Pressure and Displacements

The horizontal stress as measured by the pressure cells and fill elevation versus date at Station 306+25 are shown on Figure 2. Cell PC 1-1 has soil placed against it while tire shreds are placed against the remaining cells. It is seen that the pressure recorded by cell PC1-1 increased from zero to about 17 kPa as the first few lifts of fill were placed over the cell. This is probably in response to horizontal stress created by compaction. The stresses gradually increased as the fill elevation increased reaching a stress of between 42 and 52 kPa by the time the fill reached its final elevation of 26.9 m.

For the cells with tire shreds placed as backfill, the pressure increased by 8 to 10 kPa in response to placement of the first few lifts of tire shreds over the cell. The pressure at the lowest cell with shreds against it (PC1-2) increased to about 18 kPa as additional fill was placed. Cells PC1-3 and PC1-4 experienced little increase in stress as additional fill was placed, remaining between 8 and 10 kPa. The final stress produced by the tire shreds is less than half of the value for the cell with soil placed against it. Similar behavior was measured at Station 306+40. Thus, the design objective of using a compressible layer of tire shreds to reduce the pressure was achieved.

The measured stresses were compared to values expected based on an equivalent fluid unit weight of soil of 0.96 Mg/m^3 for at-rest conditions and 0.58 Mg/m^3 for active conditions. Cell PC1-1 is at El. 18.5 m. The overlying road surface at this station is about El. 26.8 m. PC1-1 has soil placed against it, so

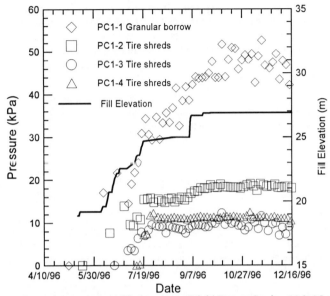

Figure 2. Lateral pressure and fill elevation at Rigid Frame, Station 306+25.

at-rest conditions are applicable. This yields a computed horizontal stress of 76 kPa compared to a measured value between 42 and 52 kPa. Cell PC1-2 is at El. 19.3 m and has tire shreds placed against it so active conditions are applicable. This yields a computed horizontal stress of 41 kPa compared to a measured value of about 18 kPa. Thus, the measured horizontal stresses for both soil and tire shreds is significantly less than the values expected based on equivalent fluid unit weights typically used for design. Possible reasons are that the equivalent fluid unit weights do not take into account the lower unit weight of the tire shreds and the effect of the approach slab on the stress distribution. Moreover, the equivalent fluid unit weights are probably somewhat conservative for the well graded granular borrow used as backfill on this project. These reasons will be investigated further as the study continues.

Horizontal movement recorded by the soil strain meters shows that the tire shred zone compressed between 10 and 30 mm. The slope indicator located 1.1 m from the interface between the tire shreds and granular borrow showed between 18 and 40 mm of movement toward the wall. In comparison, Bowles (1996) states that a wall rotation of 0.001 to 0.002H is required to achieve active conditions for dense granular backfill. This translates into a movement of 6 to 13 mm for a 6.7 m high wall. Thus, the magnitude of recorded movements are consistent with that expected for active conditions to develop.

Temperature in Tire Shred Inclusion

Thermistors were installed to monitor the temperature of the tire shred layer because of heating problems with thick tire shred fills in Washington State and Colorado (Humphrey, 1996). The temperature in the tire shreds peaked at 24 to 28°C (75 to 83°F) in August, 1996, in comparison to a temperature of 18°C (65°F) in the soil (Cosgrove and Humphrey, 1998).

NORTH ABUTMENT APPROACH FILL

The key element of the Topsham Brunswick Bypass Project was the 300-m long Merrymeeting Bridge over the Androscoggin River. The subsurface profile at the location of the North Abutment consisted of 3 to 6 m of marine silty sand overlying 13.7 to 15.2 m of marine silty clay. The clay is underlain by 6 m of glacial till and then bedrock. The existing riverbank had a factor of safety against a deep seated slope failure that was near one. The design called for an approach fill leading up to the bridge abutment that would have further lowered the factor of safety. Thus, it was necessary to devise a strategy to both improve the existing factor of safety and allow construction of the approach fill. The best solution was to excavate some of the existing riverbank and replace it with a 4.3-m thick layer of tire shreds. Tire shreds had the added advantage of reducing lateral pressures against the abutment wall. Other types of lightweight fill were considered including geofoam and expanded shale aggregate. However, tire shreds proved to be the lowest cost solution. The project used some 400,000 scrap tires.

Project Layout and Construction

The surficial marine sand was excavated to elevation 5.2 m and then the abutment wall was constructed. The north abutment is 6 m high and 24 m wide with two 4.9-m wide wing walls. The abutment is supported on steel H-piles driven to penetration resistance in glacial till or on bedrock.

A 4.3-m thick zone of tire shreds was placed from station 53+50.6 to the face of the abutment wall at station 53+72.0. The fill tapers from a thickness of 4.3 m at station 53+50.6 to zero thickness at station 53+35.4 to provide a gradual transition between the tire shred layer and the conventional fill. It was estimated that the tire shred layer would compress 460 mm due to the weight of overlying soil layers. As a result, the layer was built up an additional 460 mm so that the final compressed thickness would be 4.3 m. The tire shred layer was enclosed in a woven geotextile (Niolon Mirafi 500X) to prevent infiltration of surrounding soil. The tire shreds were spread with front end loaders and bulldozers and then compacted by six passes of a smooth drum vibratory roller (Bomag BW201AD) with a static weight of 9,432 kg. The thickness of a compacted lift was limited to 305 mm. It was determined that to obtain a compacted thickness of 305 mm, approximately 381 mm of loose tire shreds needed to be initially placed. Tire shred placement began on September 25, 1996 and was completed on October 3, 1996. A longitudinal cross-section of the completed abutment and embankment is shown in Figure 3.

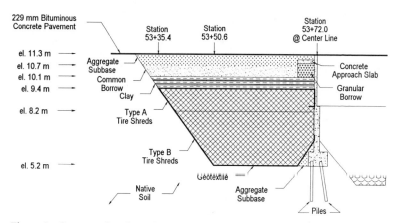

Figure 3. Cross section through North Abutment tire shred fill.

Under some conditions thick tire shred fills (greater than 8 m thick) have undergone a self-heating reaction (Humphrey, 1996).   Several factors are thought to create conditions favorable for this reaction including: small tire shred pieces, excessive exposed steel belt, presence of organic matter, and free access of air and water to the tire shred layer.   For this project, steps were taken to minimize the presence of these factors. The first was to use larger size shreds (called Type B shreds) in the lower portion of the fill from elevation 5.2 m to elevation 8.2 m. The Type B shreds had a maximum dimension measured in any direction, of 305 mm a minimum of 75% (by weight) passing the 203-mm square mesh sieve, a maximum of 25% (by weight) passing the 38-mm square mesh sieve, and a maximum of 5% (by weight) passing the 4.75-mm (No. 4) sieve.   In addition, at least one side wall was severed from the tread of each tire. Type A shreds, with the same gradation as used for the rigid frame, were placed from elevation 8.2 m to the top of the tire shred fill. It would have been preferable to use the larger Type B shreds for the entire thickness, however, a significant quantity of Type A shreds had already been stockpiled near the project prior to the decision to use larger shreds. It was judged that it would be acceptable to use the smaller Type A shreds in the upper portion of the fill.

As an additional step to reduce the possibility of self-heating, the tire shreds are overlain by a layer of compacted clayey soil with a minimum of 30% passing the 0.075 mm (No. 200) sieve. The purpose of the clay layer is to minimize the flow of water and air though the tire shreds. The clay layer is approximately 610 mm thick and is built up in the center to promote drainage toward the side slopes. A 610-mm thick layer of common borrow was placed over the clay layer. Overlying the common borrow is 762 mm of aggregate subbase.

Tire shreds undergo a small amount of time dependent settlement. For this project a thick tire shred fill adjoined a pile supported bridge abutment. This lead to concerns that there could be differential settlement at the junction with the abutment. However, Tweedie, et al. (1998) showed that most of the time dependent settlement occurs within the first 60 days. To accommodate the time dependant settlement prior to paving, the contractor was required to place an additional 305-mm of subbase aggregate as a surcharge to be left in place for a minimum of 60 days. In fact, the overall construction schedule allowed the contractor to leave the surcharge in place from October, 1996 through October, 1997. The surcharge was removed in October, 1997 and the roadway was topped with 229 mm of bituminous pavement. The highway was opened to traffic on November 11, 1997. Additional construction information is given in Cosgrove and Humphrey (1998).

Instrumentation

Four types of instruments were installed: pressure cells cast into the back face of the abutment wall; and vibrating wire settlement gauges, settlement plates and temperature sensors placed in the tire shred fill. Vibrating wire pressure cells were installed to monitor lateral earth pressure against the abutment wall. Three model TPC pressure cells (PC1-1, PC1-2, PC1-3) were installed on the face of the abutment wall 4 m right of centerline at elevations 6.7, 7.8, and 8.8 m. Three model EPC pressure cells (PC2-1, PC2-2, PC2-3) were installed 4 m left of centerline at the same elevations. Tire shreds were placed against all the cells.

Measured Horizontal Pressure and Settlement

The lateral pressure at the completion of tire shred placement (10/3/96) and completion of soil cover and surcharge placement (10/9/96) is summarized in Table 1. Lateral pressures on 10/31/96 are also shown. It is seen that at completion of tire shred placement, the pressures increased as the elevation of the cell decreased. However, at completion of soil cover and surcharge placement, the pressures recorded by cells PC1-1, PC1-2, and PC1-3 were nearly constant with depth and ranged between 17.05 and 19.61 kPa. These findings are consistent with at-rest conditions measured on an earlier project (Tweedie, et al., 1998a,b). Cells PC2-1, PC2-2, and PC2-3 showed different behavior. On 10/9/96, cell PC2-2 showed a pressure of 30.22 kPa while cell PC2-1, located only 1.07 m lower,

Table 1. Summary of Lateral Pressures on Abutment Wall.

Date	PC1-1	PC2-1	PC1-2	PC2-2	PC1-3	PC2-3
	Cell elev. = 6.70 m		Cell elev. = 7.77m		Cell elev. = 8.84 m	
10/3/96[2]	7.84[1]	7.41	6.04	7.27	2.62	1.41
10/9/96[3]	17.04	20.04	19.61	30.22	17.05	10.91
10/31/96	18.27	21.05	20.98	32.84	20.24	12.31

[1]Horizontal pressure in kPa.
[2]Date tire shred placement completed.
[3]Date soil cover and surcharge placement completed.

was 20.04 kPa and cell PC2-3, located 1.07 m above PC2-2, was 12.31 kPa. These cells were the less stiff EPC cells. Large scatter has been observed with EPC cells on an earlier tire shred project (Tweedie, et al., 1998a,b). It is thought to be due, at least in part, to large tire shreds creating a nonuniform stress distribution on the face of the pressure cell. The average pressure recorded by the three PC2 cells was 20.85 kPa which is slightly higher than the PC1 cells. Between 10/9/96 and 10/31/96 the lateral pressure increased by 1 to 2 kPa. The pressures have been approximately constant since that time.

The tire shred fill compressed about 370 mm during placement of the overlying soil cover. In the next 60 days the fill settled an additional 135 mm. Between December 15, 1996 and December 31, 1997 the fill underwent an additional 15 mm of time dependent settlement. The rate of settlement had decreased to a negligible level by late 1997. The total compression of the tire shred fill was 520 mm which was 13% greater than the 460 mm that was anticipated based on laboratory compression tests. The final compressed density of the tire shreds was about 0.9 Mg/m³.

Temperature of Tire Shred Layer

A small amount of self heating of the tire shreds occurred. Five out of the 12 thermistors in Type A shreds experienced a peak temperature of between 30 and 40°C (86 and 104°F). In contrast, only two of the 18 thermistors in the larger Type B shreds experienced a peak in this range and these two sensors may have been influenced by warmer overlying Type A shreds. This suggests that larger shreds are less susceptible to heating. In any case, the peak temperatures were well below those needed to initiate combustion of tires. Since early 1997, the overall trend has been one of decreasing temperature, however, the temperature of the shreds do appear to be slightly influenced by seasonal temperature changes.

JETPORT INTERCHANGE

The Jetport Interchange is being constructed on the Maine Turnpike in Portland, Maine to provide more direct access to the Turnpike from the Portland Jetport and surrounding major roadways. The Interchange consists of a new connector roadway with a bridge over the Maine Turnpike, northbound and southbound on/off ramps and associated toll plaza. The connector extends from the Jetport at Johnson Road in Portland, over the Maine Turnpike on a new bridge approximately 162 m south of the existing Congress Street overpass, and ends at the intersection of Congress Street and Hutchins Drive in Westbrook. The new bridge will be approximately 111 m long with integral, pile-supported abutments at each end, and two intermediate piers. Approach embankments, up to 9.8 m high, have been constructed at each abutment. Due to the height of the approach embankments and the consistency of the foundations soils, soft ground construction techniques were required to provide acceptable slope stability and limit post-construction settlement. To meet the project schedule, the approach embankments were constructed under an early embankment preload contract.

The highway portions of the Interchange were designed by Wright-Pierce Engineers of Topsham, Maine for HNTB, Inc., of Portland, Maine, the coordinating consultant for the Maine Turnpike Authority.   Geotechnical engineering services were provided by Haley & Aldrich, Inc., of South Portland, Maine.

Subsurface Conditions

Subsurface explorations disclosed up to 12 m of marine clay underlying the embankments. Laboratory testing was undertaken to determine the compressibility parameters, stress history and undrained strength.   The testing program included one-dimensional   consolidation   tests,   unconsolidated-undrained   triaxial compression tests and classification testing consisting of Atterberg limits, shear vane and natural water content.   Undrained shear strength was also measured by field vane shear tests in the borings.   The test results indicate that the clay is an overconsolidated, moderately sensitive, inorganic clay of low plasticity.   Undrained shear strength varied from approximately 72 kPa near the top to 19 kPa near the center of the layer.

Slope Stability and Settlement

Initial studies indicated that the 9.8-m high embankments would have unacceptable factors of safety for slope stability and post construction settlement (including one cycle of secondary compression) on the order of 380 to 460 mm were estimated.   To increase factors of safety, alternatives such as: 1) stabilizing berms; 2) ground improvement such as stone columns and deep soil mixing; and 3) lightweight fill were considered.   Stabilizing berms were eliminated due to their impact on adjacent wetlands.   Lightweight fill was selected as the best remaining alternative based on lower construction cost and tire shreds proved to be the lowest cost lightweight fill.

Slope stability analysis indicated that up to 6.1 m of tire shreds would be required for stability.   But tire shreds were not enough.   In order to begin construction of the bridge within twelve months after the start of embankment construction the strength of the clay had to be increased and the post-construction settlement had to be reduced.   Vertical drains and surcharging were the answer. Prefabricated vertical drains spaced 1.83 m on center would allow the clay to consolidate and gain sufficient strength to yield acceptable factors of safety at the end of construction.   The addition of a surcharge for an approximate eight month period would reduce post construction settlements to acceptable levels.

Typical Embankment Section

As noted above some thick tire shred fills (greater than 8 m thick) have undergone a self-heating reaction.   Two of these projects were located in Washington State and one was in Colorado.   These projects were constructed in 1995 and each experienced a serious self heating reaction within 6 months after completion (Humphrey, 1996).   The lessons learned from these projects were condensed into design guidelines developed by the Ad Hoc Civil Engineering Committee (1997), a partnership of government and industry dealing with reuse of scrap tires for civil engineering purposes. For tire shred layers ranging in thickness from 1 to 3 m, the guidelines give the following recommendations:

- Tire shreds shall be free of contaminants such as oil, grease, gasoline, diesel fuel, etc., that could create a fire hazard
- In no case shall the tire shreds contain the remains of tires that have been subjected to a fire
- Tire shreds shall have a maximum of 25% (by weight) passing 38-mm sieve
- Tire shreds shall have a maximum of 1% (by weight) passing 4.75-mm sieve
- Tire shreds shall be free from fragments of wood, wood chips, and other fibrous organic matter
- Tire shreds shall have less than 1% (by weight) of metal fragments that are not at least partially encased in rubber
- Metal fragments that are partially encased in rubber shall protrude no more than 25 mm from the cut edge of the tire shred on 75% of the pieces and no more than 50 mm on 100% of the pieces
- Infiltration of water into the tire shred fill shall be minimized
- Infiltration of air into the tire shred fill shall be minimized
- No direct contact between tire shreds and soil containing organic matter, such as topsoil
- Tire shreds should be separated from the surround soil with a geotextile
- Use of drainage features located at the bottom of the fill that could provide free access to air should be avoided

The guidelines further recommend that the maximum thickness of a tire shred layer be 3 m. The guidelines also give less stringent requirements for tire shred layers less than 1 m thick. The Portland Jetport Project is the first major tire shred fill that was designed according to the guidelines.

The Portland Jetport Project required up to 6.1 m of tire shreds, so the tire shreds were separated into two layers. The upper layer was maintained at a uniform thickness of 3.05 m and the lower layer thickness varied from 0.61 to 3.05 m. To account for compressibility of the tire shreds as fill was added to the embankment, the lower tire shred layer was over-built by 15 percent and the upper layer was over-built by 10 percent.

Several design details were incorporated into the embankment construction to reduce the potential for self-heating of the tire shreds. Type B shreds were specified. They had the same gradation requirements as for the North Abutment except that only 1 percent by weight passing the 4.75-mm sieve was allowed. In addition, the tire shreds were required to have less than 1 percent by weight of metal fragments which were not at least partially encased in rubber. The tire shred layers were wrapped in woven geotextile filter fabric to act as a separator and prevent migration of soil into the tire shred layer. Each tire shred layer was encapsulated in inorganic, low permeability soil to minimize water infiltration and air flow into the shred layer. The top of the low permeability soil layers were crowned and a well graded gravel drainage layer was provided at the base or side of each tire shred layer to allow any water present to drain to the side slopes. A typical cross section is shown in Figure 4.

Instrumentation

Instrumentation installed to monitor the performance of the embankments included the following: temperature sensors to measure the temperature of the tire shreds in the upper and lower layers; piezometers to measure the pore water pressure and rate of dissipation in the clay stratum; settlement platforms to measure the settlement of the clay stratum and the two tire shred layers; and inclinometers to measure lateral movement in the foundation soils and embankment.

In general, the thermistors (temperature sensors) were placed in the center of the layer at the centerline and 7.3 m left and right of centerline at 7.6 m intervals in the lower tire shred layers and at the centerline and 3.7 m left and right of centerline at 7.6 m intervals in the upper tire shred layer. A total of 39 thermistors were installed. The vibrating wire piezometers were placed in groups of three at six separate locations (three in each embankment), with each piezometer placed at the quarter points of the clay thickness. A total of 18 piezometers were installed.

Settlement platforms were installed in groups of three, one at the top of the clay and one each at the top of the tire shred layers to measure independently the settlement of the clay layer and the compression of each tire shred layer. Platforms consisted of a 610-mm square plywood base and a 51 mm diameter black iron pipe riser. A 102-mm diameter PVC pipe was placed around the riser pipes of the two lower platforms to protect against drag forces due to compression of the shred layers. A total of 26 settlement platforms were installed in the two embankments.

Seven inclinometers were installed at the two embankments; five were installed at the toe of the slope to monitor lateral movement of the foundation soils and two were installed through the embankments to measure lateral movement of the tire shred layers in addition to movement of the foundation soils.

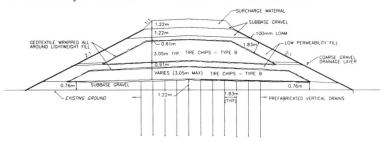

Figure 4. Typical embankment section, Portland Jetport

Tire Shred Placement

Tires were shredded off-site at a tire stockpile in Durham, Maine and trucked to the site for placement at a rate not to exceed 544 metric tons per day. Tire shreds were placed in maximum 305-mm thick layers using a track-mounted bulldozer and compacted with a minimum of six passes of a vibratory smooth drum roller weighing at least 9,072 kg. Embankment construction began in mid-June 1997 and was completed with the surcharge in place by 30 September 1997. A total of approximately 10,430 metric tons of tire shreds (estimated to be 1,200,000 tires) were placed in the two embankments. Based on the volume of the tire shred layers and the weight of shreds delivered to the project, the average in-place density was calculated to be 0.78 Mg/m^3. This is lower than the density for the North Abutment Fill and is possibly due to the larger size of the tire shreds.

Embankment Performance

Temperature sensors showed an initial high temperature due to exposure to sunlight and warm air temperatures with gradually decreasing temperatures following encapsulation. Present temperatures range from a low of 12°C (52°F) to a high of 23°C (73°F). It is anticipated that temperatures will continue to decrease through early summer and then possibly increase slightly in early to mid-fall in response to warmer seasonal temperatures.

Piezometers disclosed that pore pressures increased as the embankment was constructed and then decrease with time under constant load with the greatest increase in pore pressure occurring where the embankment fill was the greatest.

Settlement platforms indicate that, as of March, 1998, much of the clay stratum continues to exhibit time-dependent settlement as pore pressures decrease and water is squeezed from the clay. Log time plots suggest that the clay has completed primary consolidation in areas of least clay thickness, but for most of the clay, primary consolidation is still ongoing. Clay settlement after approximately 5 months at full load varies from approximately 50 to 380 mm. Settlement data for the tire shred layers indicates rapid compression during loading with very little time-dependent movement. Settlement of the upper shred layer after approximately 5 months at full load varies from 125 to 530 mm. Settlement of the lower shred layer varies from 125 to 460 mm.

The inclinometer data shows no significant movement in the foundation soils, with the greatest lateral movement on the order of 13 mm after 5 months at full load. Inclinometers through the tire shred layers show erratic movement during construction, likely due to unbalanced loading during compaction around the inclinometers. The maximum lateral movement following completion of the shred layers is on the order of 8 mm after 5 months at full load.

It is significant to note that all instruments appear to be working correctly and no apparent damage occurred during construction.

SUMMARY

These projects demonstrate that tire shreds can be used as lightweight backfill for retaining walls and lightweight embankment fills. Advantages include: reduced lateral pressure, increased stability, reduced settlement, and reduced overall construction cost. Moreover, these projects show that properly designed tire shred fills do not experience a deleterious self heating reaction.

Acknowledgments

The authors thank the Maine Department of Transportation, Maine Turnpike Authority, T.Y. Lin International, and HNTB, Inc. for their support throughout these projects. CPM Constructors, H.E. Sargent, Reed and Reed, Harry C. Crooker and Sons, Drainage & Ground Improvement, Inc., Maine Test Borings, Inc., AD Electric, and Enterprise Electric are thanked for their careful job constructing the projects and installing instrumentation. The tire shred suppliers, J.P. Routhier and Sons, Arthur Schofield, Inc., Sawyer Environmental Recovery, and Casella Tires, are thanked for providing a high quality tire shreds.

References

Ad Hoc Civil Engineering Committee (1997), "Design Guidelines to Minimize Internal Heating of Tire Shred Fills," Scrap Tire Management Council, Washington, D.C., 4 pp.

Bowles, J.E. (1996), Foundation Analysis and Design, 5th Edition, McGraw-Hill, New York.

Cosgrove, T.A., and Humphrey, D.N. (1998), "Monitoring Program for Two Tire Shred Fills on the Topsham-Brunswick Bypass Project," a report to Maine Dept. of Transportation by Dept. of Civil and Environmental Engineering, University of Maine, Orono, Maine.

Humphrey, D.N. (1996), "Investigation of Exothermic Reaction in Tire Shred Fill Located on SR 100 in Ilwaco, Washington," report to Federal Highway Administration, Washington, D.C., 61 pp.

Humphrey, D.N., Cosgrove, T., Whetten, N.L., and Hebert, R. (1997a), "Tire Chips Reduce lateral Earth Pressure Against the Walls of a Rigid Frame Bridge," Maine Section ASCE Technical Seminar, Portland, Maine, 11 pp.

Humphrey, D.N., Katz, L.E., and Blumenthal, M. (1997a), "Water Quality Effects of Tire Chip Fills Placed Above the Groundwater Table," Testing Soil Mixed with Waste or Recycled Materials, ASTM STP 1275, M.A. Wasemiller and K.B. Hoddinott, Eds., American Society for Testing and Materials, pp. 299-313.

Tweedie, J.J., Humphrey, D.N., and Sandford, T.C. (1998b), "Tire Shreds as Lightweight Backfill for Retaining Walls - Phase II," a report to New England Transportation Consortium by Dept. of Civil and Environmental Engineering, University of Maine, Orono, Maine, 314 pp.

Tweedie, J.J., Humphrey, D.N., and Sandford, T.C. (1998b), "Full Scale Field Trials of Tire Chips as Lightweight Retaining Wall Backfill, At-Rest Conditions", Transportation Research Record, Transportation Research Board, Washington, D.C.

Geotechnical Performance of a Highway Embankment Constructed Using
Waste Foundry Sand

David G. Mast[1], A.M. ASCE, and
Patrick J. Fox[2], A.M. ASCE

ABSTRACT

A highway embankment was constructed near Auburn, Indiana, using waste foundry sand (WFS) as a demonstration project for the Indiana Department of Transportation (INDOT). Laboratory testing before construction indicated the WFS is
best characterized as a silty sand, with good strength properties and a relatively
low hydraulic conductivity. Field instruments were installed in three embankment
sections composed of WFS, clay, and clean sand. Instrumentation for the WFS
section consisted of vertical and horizontal inclinometers, settlement plates, total
pressure cells, and piezometers. Measurements up to 21 months after WFS
placement showed small vertical and horizontal deformations. Comparison of
compaction testing methods for WFS indicated some corrections may be necessary
when using a nuclear density gauge. Field and laboratory tests indicated the WFS
had a hydraulic conductivity on the order of 10^{-7} m/s. Standard penetration testing
of the completed embankments showed blow counts for the WFS were comparable
to those for the clean sand. The economic benefits gained by the foundry were estimated to be $640,000, possibly higher when considering siting and construction
costs for a new landfill. Savings in material costs for INDOT were estimated to be
$145,000 as a result of using WFS for this project.

INTRODUCTION

In Indiana, about 450,000 tons of waste foundry sand (WFS) are landfilled annually (INCMA 1992). The estimated disposal cost of this material is $15 million for
Indiana foundries (Mast 1997). Thus, there is considerable interest on behalf of

[1] Staff Engineer, NTH Consultants, Ltd., 277 Gratiot, Suite 600, Detroit, MI
48226. Phone: (313) 965-0036. Email: mast@wwnet.net

[2] Associate Professor, School of Civil Engineering, Purdue University, West
Lafayette, IN 47907. Phone: (765) 494-0697. Email: pfox@ecn.purdue.edu.

foundries to find alternate uses for WFS in cases where the material poses no threat to the environment. This paper presents the results of a field demonstration project in which WFS was used to construct a highway embankment for the Indiana Department of Transportation (INDOT). Testing and monitoring for this project provide evidence that, from a geotechnical standpoint, WFS can be successfully utilized in place of natural fill materials. Information regarding the environmental performance of the WFS is presented in Partridge et al. (1998).

Considerable research into waste material reuse has been initiated by INDOT and the Indiana Cast Metals Association (INCMA) in response to legislation passed in 1991 by the Indiana General Assembly. This legislation encourages waste material reuse in construction by exempting certain Type III waste materials from full coverage under Resource Conservation and Recovery Act (RCRA) regulations. In Indiana, Type I, II, or III materials normally must be landfilled in an approved site. However, the new legislation provides for the "...beneficial reuse of foundry sand meeting the Type III category, if the construction reuse is 'legitimate', including the use as pavement base".

WFS PRODUCTION / ORIGIN

Waste foundry sand is a byproduct of the metal casting process. The WFS used in this project originated from the Auburn Foundry, Inc. of Auburn, Indiana. Auburn Foundry uses a green sand molding process in which molten iron is poured into sand molds while the sand is still damp from the mold forming procedure. The major components of green sand are mixed in the following proportions (Fox et al. 1997):

- 85 to 95% uniform quartz sand,
- 4 to 10% bentonite clay to provide cohesion,
- 2 to 10% combustible additives (e.g., sea coal) to provide a surface finish on the castings,
- 0.5 to 5% iron oxide to increase mold strength, and
- 2 to 5% water.

After the iron has cooled sufficiently, the green sand is shaken off the castings, and the majority (about 90%) is mixed with fresh materials and returned to production. The leftover sand, along with other waste materials, is sent to the Auburn Foundry private landfill (monofill) as WFS. The other waste materials may include intact or crushed sand cores, "popcorn slag", dust collected around the conveyor systems and shakeout chambers, defective iron castings, and miscellaneous items such as discarded welding rods. Sand cores are used to create void spaces or deep impressions inside the final castings, and are prepared separately by a chemical-bonding or shell-bonding process. Popcorn slag is a lightweight, glassy material that is formed when iron slag from the cupola melting process comes in contact with moisture.

GEOTECHNICAL LABORATORY TESTING

Geotechnical laboratory tests were performed on "fresh" and "weathered" samples of Auburn Foundry WFS. The fresh WFS sample was obtained immediately after delivery to the monofill from the foundry plant recycling system. The weathered WFS samples were taken from the top of the Auburn Foundry monofill and had been exposed to weather conditions for some time. Table 1 presents a summary of geotechnical laboratory test results for the Auburn Foundry WFS samples.

Figure 1 shows percent finer versus particle diameter for fresh and weathered Auburn Foundry WFS. The plot also includes data from several Indiana green sand samples tested by Javed and Lovell (1994) and tests performed by the INDOT materials laboratory on Auburn WFS. The materials tested by Javed and Lovell and by INDOT, which were not specified as fresh or weathered, had a fines content ranging from 6 to 38%. The Auburn Foundry fresh WFS had 40% fines and the weathered samples had 10 to 22% fines. The lower fines content for the weathered samples was likely due to washing of fines from the surface to the interior of the monofill by rain.

The specific gravity (G_s) of Auburn WFS was approximately 2.53, which is lower than expected for a material composed primarily of quartz sand. However, similar results were obtained from previous laboratory tests of WFS in which seven Indiana green sands had an average G_s of 2.49 (Javed et al. 1994).

Laboratory Proctor compaction tests were used to evaluate the maximum dry unit weight ($\gamma_{d, max}$) and the optimum moisture content (OMC) for weathered and fresh WFS. The majority of the samples tested were collected from the top of the monofill, with a few points added after completion of the test pad (discussed later).

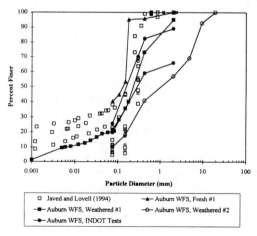

Figure 1: Particle Size Distributions for WFS

The weathered WFS samples were tested using both the standard and modified Proctor Methods, utilizing the Method B procedure. The Method B procedure is recommended for materials with more than 20% retained on the No. 4 sieve (4.75 mm) and 20% or less retained on the ⅜-in. sieve (9.5 mm). Method B allows particles passing the ⅜-in. (9.5 mm) sieve to

Table 1: Geotechnical Laboratory Test Results for Auburn Foundry WFS

Test		Symbol	Weathered WFS	Fresh WFS
Liquid Limit (ASTM D-4318)		w_L %	30.7	--
Plastic Limit (ASTM D-4318)		w_P %	24.7	--
Specific Gravity of Solids (ASTM D-854)		G_s	2.53	2.46
Percentage of Coarse Particles (ASTM D-422)			78–90%	60%
Percentage of Fines (Passing # 200 Sieve) (ASTM D-422)		P200	10 - 22%	40%
Percentage of Clay Size Particles (< 0.005 mm) (ASTM D-422)			9%	--
Standard Proctor Compaction Test, Method B (ASTM D698)	Optimum Moisture Content	OMC	15.5%	--
	Maximum Dry Unit Weight	$\gamma_{d,\,max}$	16.8 kN/m^3	--
Modified Proctor Compaction Test, Method B (ASTM D1557)	Optimum Moisture Content	OMC	12.8%	--
	Maximum Dry Unit Weight	$\gamma_{d,\,max}$	18.2 kN/m^3	--
Vibratory Compaction Test (ASTM D-4253)	Optimum Moisture Content	OMC	28.8%	--
	Maximum Dry Unit Weight	$\gamma_{d,\,max}$	13.3 kN/m^3	--
Direct Shear Test (ASTM D-3080)	Cohesion Intercept, Loose Specimen	c'	6.9 kN/m^2	13.8 kN/m^2
	Cohesion Intercept, Dense Specimen	c'	13.1 kN/m^2	15.2 kN/m^2
	Internal Friction Angle, Loose Specimen	ϕ'	35°	33°
	Internal Friction Angle, Dense Specimen	ϕ'	38°	39°
California Bearing Ratio Test (ASTM D-1883)	Soaked Specimen (5 mm penetration)	CBR	16.8	6.2
	Unsoaked Specimen (5 mm penetration)	CBR	17.9	14.1
CBR Swell Test	% Swelling		0 %	0.9 %
Flexible-Wall Hydraulic Conductivity Test (ASTM D-5084)	Standard Proctor, Wet of OMC	k	1.2 x 10^{-8} m / s	--
	Modified Proctor, Dry of OMC		7.1 x 10^{-7} m / s	

be included in the sample. Using this criterion, the Auburn WFS generally should be tested using the Method B procedure. Similar to natural soils, OMC decreased and $\gamma_{d,\ max}$ increased with increasing compaction energy. Limited testing of fresh and weathered WFS using the more common Method A procedure (i.e., material passing the No 4 sieve) resulted in lower $\gamma_{d,\ max}$ and higher OMC than the Method B procedure. A vibratory compaction test on weathered WFS indicated that vibration was not particularly effective in compacting this material.

For fresh WFS, California Bearing Ratio (CBR) values were 56% less for 4-day soaked specimens than for unsoaked specimens. The same comparison for weathered WFS showed only a 6% drop in CBR value for soaked and unsoaked specimens. This indicates the significance of fines content on the CBR of WFS. Fresh WFS showed less than 1% volumetric swelling, whereas weathered WFS showed no measurable swelling.

Laboratory hydraulic conductivity tests for weathered WFS were performed on backsaturated 102-mm diameter specimens in a flexible-wall permeameter. The measured hydraulic conductivity values were 1.2×10^{-8} m/s for the standard Proctor, wet of OMC specimen and 7.1×10^{-7} m/s for the modified Proctor, dry of OMC specimen. These two tests suggest that, similar to natural soils, the hydraulic conductivity of weathered WFS may be more sensitive to compaction moisture content than to compaction energy.

Direct shear tests on dry WFS samples showed that the material had a small cohesive intercept, which was smaller for weathered than fresh specimens (presumably due to reduction in fines content). Strength properties for the dense WFS are in good agreement with values measured by Javed et al. (1994).

PROJECT DESCRIPTION

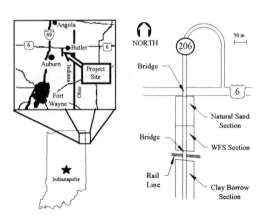

Figure 2: Project Location and Plan View

The demonstration project involved a 244 m long section of County Route (C.R.) 206 near Auburn, Indiana (Figure 2). The roadway was designed to cross over an active rail line 229 m south of S.R. 6 and then cross over S.R. 6 itself, intersecting from the north after a gradual change in grade. The embankment between the bridges is 9.1 m high, 73 m wide at the base, and has 3H:1V side slopes. The other bridge approach embankments were of the same dimensions. The C.R. 206 embankments were well-suited for the proposed research. First, the Auburn Foundry showed interest in providing material for the project and also maintained a large stockpile of WFS in their own monofill. Secondly, INDOT agreed to build one-half of the embankment between the bridges using a clean natural sand (classified as B-borrow), and the approach embankments with clay borrow. The use of three different materials allowed us to compare their performance during and after construction of the embankments.

The site is underlain by shale and limestone bedrock at a depth of approximately 67 m. The soil above bedrock are primarily glacial in origin, including both ground moraine and end moraine deposits. The original topography of the site was flat, with several wetlands in the area. Soil borings obtained by INDOT prior to construction revealed the following general soil profile under the site: medium stiff loam, from 0.3 to 0.6 m below ground surface; loose sand from 0.6 to 1.5 m; medium dense sand from 2.4 to 3.7 m; very stiff to hard clay from 3.7 to 8.2 m; medium dense sand and gravel from 8.2 to 11 m. The groundwater table was encountered at approximately 4.3 m during drilling, and was measured at a depth of 1.7 m 24 hours later.

FIELD INSTRUMENTATION

Figure 3 presents a schematic diagram of the geotechnical instrumentation for the WFS and clean sand (B-borrow) sections of the embankment. The diagram also shows groundwater monitoring wells and lysimeters used for the environmental investigation. Two vertical inclinometers were installed prior to construction to monitor lateral movements in the WFS section of the embankment. Readings were taken weekly during construction and at increasing intervals thereafter. Additional inclinometer tubing was added as the embankment height rose in order to allow access for monitoring.

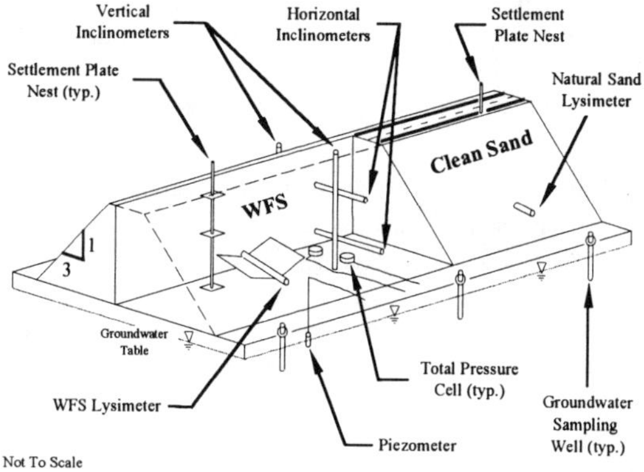

Two horizontal inclinometers were installed in the WFS portion of the embankment to measure settlement of the foundation soils and vertical compression of the WFS. The lower horizontal inclinometer was installed at the base of the embankment. The upper horizontal inclinometer was installed at a height of

Figure 3: Embankment Instrumentation

5.5 m above the base of the embankment, approximately 2/3 of the total embankment height. The ends of each inclinometer tube were surveyed regularly to establish their elevations with time.

Four nests of settlement plates were installed during construction; two in the WFS section, one in the B-borrow section, and one in the clay borrow embankment. For each nest, square settlement plates (0.9 m x 0.9 m) were set at the base, near the mid-height, and near the top of the embankment. Each settlement plate was connected to its own 51 mm steel pipe which was then sleeved with a PVC pipe to eliminate friction with the surrounding soil. The tops of the pipes were surveyed regularly relative to a known benchmark. By plotting differences in plate elevations versus time, settlement of the foundation soils was measured, as well as net compression of the WFS between the top and base plates within the embankment.

Two pneumatic total pressure cells were installed at the base of the embankment in the WFS section. One pressure cell was installed underneath the centerline (CL) of the embankment and the other was installed underneath the shoulder of C.R. 206, which was 7.0 m from centerline. Two pneumatic piezometers were installed at a depth of 6.1 m below original grade underneath the WFS section, and were located underneath each shoulder of C.R. 206.

FIELD TESTING RESULTS

Field Compaction

Two full-scale test pads were constructed at the monofill site to develop a method specification for field compaction and evaluate field unit weight testing procedures. We were concerned that the material used for the initial Proctor compaction tests (i.e., from the top of the monofill) was not representative of the material in the entire monofill. In addition, there was concern that nuclear density gauge and sand cone tests would not provide accurate unit weight test results for WFS. It was suggested that miscellaneous metal pieces, such as defective iron castings and discarded welding rods, would interfere with the nuclear density gauge test. The presence of foreign objects and the hardness of compacted WFS were expected to make sand cone unit weight testing difficult.

Material for the test pads was obtained from within the body of the stockpile using a front-end loader. At that time, WFS samples were obtained for additional laboratory compaction testing. One test pad was compacted using a vibratory smooth drum, and the other with a 36,000 kg rubber-tire bulldozer. The test procedure was to compact the pad with one roller or bulldozer pass, then measure the dry unit weight and moisture content using the nuclear density gauge and sand cone. This procedure was repeated for a total of 10 passes. One-half of each test pad was watered and then compacted with the same number of passes as the dry side to evaluate the effect of watering. The test pads showed that the rubber-tire bulldozer was more effective in compacting WFS, and that significant watering was generally not necessary to reach OMC. The vibratory action of the smooth drum roller tended to loosen the WFS layers which had been compacted on the previous pass. In contrast, the kneading action of the rubber tires produced a tightly compacted surface without peeling or cracking. As a result of these field tests, INDOT produced the following method specification for compaction of the Auburn Foundry WFS:

1. Each embankment lift of WFS shall receive a minimum of 6 passes with the pneumatic (rubber) tire bulldozer. The weight of the bulldozer should be 36,000 kg (79,500 lb.) with a tire pressure of 207 kPa (30 psi).

2. During compaction, the moisture content of the WFS shall be maintained between 12% and 15%.

3. Prior to compaction, the WFS shall be placed in 200 mm (8 in) loose lifts.

INDOT also concluded that moisture contents should be determined in the field using the Speedy Moisture Meter (SMM), with a correction factor of -1 percentage point. For cohesionless soils, the SMM provides a rapid (≤ 2 min) measurement of moisture content. Field engineers were instructed not to depend on the nuclear density gauge for quality control in the field. Instead, sand cone tests were used to verify successful compaction of the WFS in accordance with the above method specification and for comparison to nuclear density gauge unit weight tests performed for research.

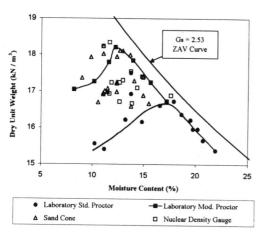

Figure 4 presents field compaction data taken during embankment construction, along with standard and modified Proctor compaction curves obtained in the laboratory. The field data generally fell between the two Proctor curves, indicating that the Method B standard and modified Proctor tests provided good indications of the unit weight of WFS when compacted according to the method specification.

Figure 4: Field and Laboratory Compaction Testing of Auburn WFS

<u>Comparison of Field Compaction Testing Methods</u>

Near the completion of the embankment, a 3.6 m square WFS test section was used to directly compare results from nuclear density gauge, sand cone, and SMM tests. Within this test section, six randomly located nuclear density gauge tests using a 150 mm (6 inch) probe depth were performed adjacent to six sand cone tests. For the sand cone tests, moisture contents were evaluated using the SMM and conventional oven drying.

Figure 5 presents a comparison of moisture contents obtained using the nuclear density gauge and SMM with oven-dry values. The dashed lines represent a 5 percentage point offset from perfect correlation. Although moisture contents obtained using the nuclear density gauge and SMM were generally lower than those obtained by oven drying, the SMM was more reliable for the measurement of WFS moisture contents for this project. Future testing of WFS using a nuclear density gauge may require a correction factor for moisture content.

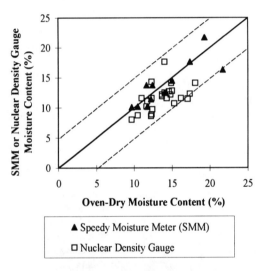

Figure 5: Field Measurements of Moisture Content of Auburn WFS

Dry unit weights measured for the test section, as well as four side-by-side field tests performed at random during construction, are shown in Figure 6. The open squares show nuclear density gauge dry unit weights plotted against sand cone dry unit weights calculated using SMM moisture contents. The solid circles show the same nuclear density gauge dry unit weights plotted against sand cone dry unit weight values calculated using oven-dry moisture contents. The triangles show the random side-by-side tests, for which sand cone dry unit weights were obtained using the SMM. The nuclear density gauge dry unit weight measurements agree well with nearly all of the sand cone results. For this project, moisture content and wet unit weight measure-ments for the nuclear density gauge did not agree closely with the SMM and sand cone values, but the dry unit weights were in good agreement.

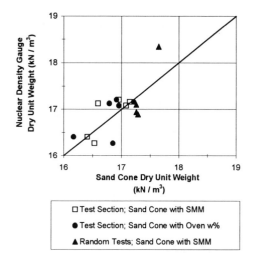

Figure 6: **Field Measurements of Dry Unit Weight of Auburn WFS**

Embankment Deformation

Figures 7 and 8 present change in elevation of the bottom and top horizontal inclinometers in the WFS section versus horizontal distance from the west side of the embankment. Each plot shows change in elevation relative to a baseline reading taken soon after installation. A negative displacement value indicates settlement. The irregular pattern in the bottom inclinometer settlement is likely a result of variations in foundation soil stiffness and applied stress across the width of the embankment. Between March 13 and October 19, 1997, it appears that the bottom inclinometer pipe became clogged. As a result, the instrument did not read properly between 0 and 20 m on October 19. Taking this error into account, settlement reached a maximum value of 45 mm. Similarly, the top inclinometer showed an initial settlement of 15 mm soon after installation and a total settlement of 25 mm after 21 months. The top inclinometer showed less settlement than the bottom inclinometer because it was installed after the majority of the WFS had been placed in the embankment.

Figures 9 through 11 present the change in elevation for settlement plates that remained undamaged during construction. The readings start at the completion of WFS placement on August 15, 1996. Although errors introduced by surveying caused some erratic readings, the measurements show a trend of increasing settlement with the same order of magnitude as the horizontal inclinometers. In the clay borrow section, the two upper settlement plates in the nest experienced up to 45 mm of settlement during 625 days of monitoring. Before being damaged at 60 days, plates for the B-borrow section showed maximum settlement of 9 mm, compared to a maximum of 5 mm and 17 mm for WFS and clay borrow over the same time period. Finally, the surviving WFS nest (designated as Foundry Sand South) showed a maximum settlement of approximately 15 mm at 625 days.

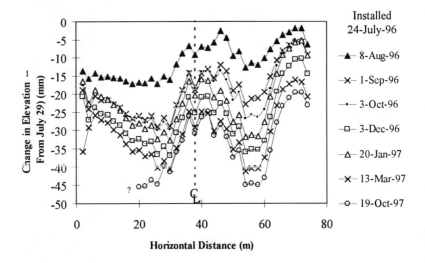

Figure 7: Settlement of Bottom Horizontal Inclinometer

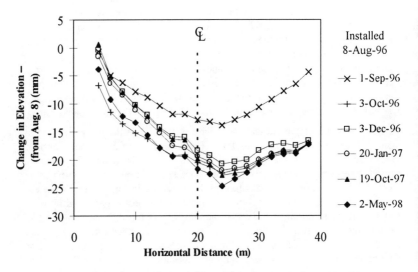

Figure 8: Settlement of Top Horizontal Inclinometer

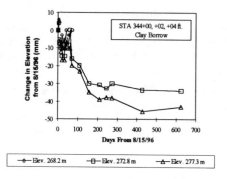

Figure 9: Settlement of Clay Borrow Settlement Plates

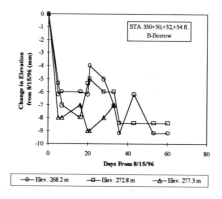

Figure 10: Settlement of B-Borrow Settlement Plates

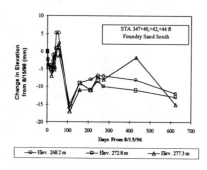

Figure 11: Settlement of Waste Foundry Sand South Settlement Plates

A more detailed analysis showed the final measured net compression between the top and bottom plates was as follows: clay borrow, 7 mm; B-borrow, 7 mm; and WFS (Foundry Sand south), 3 mm (Mast 1997). The WFS performed as well or better than the sand and clay fill materials used for this project, although the net compression based on available data was acceptable for all materials.

Figures 12 and 13 present lateral displacements calculated from vertical incli-nometer readings. Due to damage during con-struction, these readings begin after completion of the WFS embankment. Profiles for the west vertical inclinometer indicate lateral displacements of 3 to 4 mm up to March 13, 1997, with a small in-crease in October, 1997, at a depth of 4.5 m. In May and June, 1998, significant horizontal movement was recorded in the WFS section. Repeated measurements have shown the indicated movement is not due to instrument error. Continued monitoring is warranted for the west side of the embankment.

The east vertical inclinometer experienced lateral displacements as large as 5 mm in the period between the baseline reading date to March 13, 1997. The displace-ment increased to a maximum of 11 mm on October 19, 1997. Inclinometer read-ings were somewhat erratic throughout the entire monitoring period, which sug-gests small errors in the measurements.

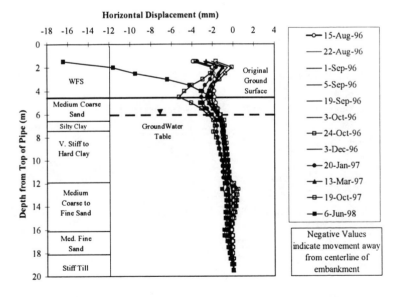

Figure 12: Horizontal Displacements of West Vertical Inclinometer

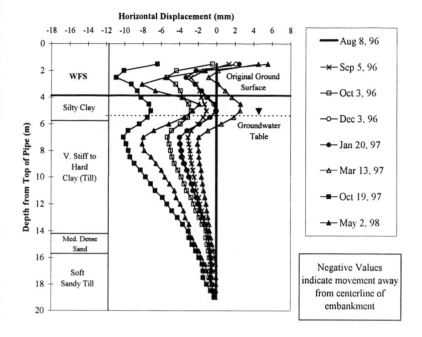

Figure 13: Horizontal Displacements of East Vertical Inclinometer

In Situ Hydraulic Conductivity Testing

In order to verify the laboratory hydraulic conductivity measurements for compacted Auburn WFS, a sealed double-ring infiltrometer (SDRI) test was performed in the field (Figure 14). The SDRI eliminates evaporation losses during testing and produces aproximately 1-dimensional flow through the inner ring (Trautwein and Boutwell 1994). Prior to SDRI installation, the hardened WFS surface material was loosened to a

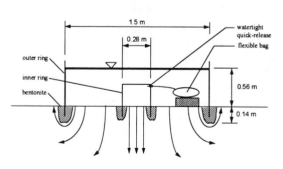

Figure 14: SDRI Test of WFS

depth of approximately 51 mm. The SDRI was then filled with water and a 3-liter flexible plastic bag was attached to the inner ring. The bag was intermittently weighed to determine inflow through the inner ring to the WFS over time. Tensiometers were not installed under the SDRI due to the hardness of the compacted WFS. Construction operations required removal of the SDRI after 18 days of inflow measurements.

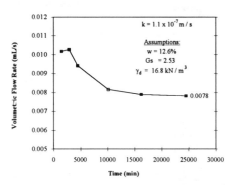

Figure 15: SDRI Volumetric Flow Rate vs. Time

Figure 15 presents volumetric flow rate versus time for the SDRI test. Prior to termination of the test, the flow rate approached a nearly steady value of 0.0078 mL/s. To calculate the position of the wetting front, it was assumed that the inflow water saturated a cylinder of WFS directly below the inner ring. The pore pressure at the wetting front was assumed to be zero. Thus, the hydraulic gradient (i) was calculated as follows:

$$i = \frac{H_w + L}{L}$$

where:

H_w = height of water in the outer ring

L = calculated depth of wetting front.

Based on the WFS properties given in Figure 15, the initial porosity and initial degree of saturation were 0.32 and 57%, respectively. The total inflow included 12,190 mL from bag readings and 3.5 days of saturation prior to the first reading at an assumed flowrate of 0.010 mL/s. Knowing the total volume of inflow water (15,210 mL) and the inner ring area (0.0616 m²), the depth of the wetting front is calculated as 2.34 m. The height of water in the outer ring remained nearly constant throughout the test at 0.475 m. Thus, the hydraulic gradient at the end of the test was 1.20 and the hydraulic conductivity for the compacted WFS is estimated as 1.1×10^{-7} m/s. These field results are consistent with field observations of low hydraulic conductivity made in a similar WFS embankment project in Wisconsin (Lovejoy et al. 1996). In addition, the estimated hydraulic conductivity value falls within the range of laboratory testing results presented in Table 1. An upper bound to the estimate of hydraulic conductivity can be calculated using a unit gradient ($i = 1$), which gives a value of 1.3×10^{-7} m/s.

It should be noted no movements of the inner ring were measured to obtain a swell correction for this test. It was expected this correction would be minimal because the CBR swell test showed negligible swelling for weathered WFS samples. This test indicates that the bentonite and other fines reduce the hydraulic conductivity of compacted Auburn WFS from that expected of a sandy material, as the term waste foundry "sand" implies. Compacted Auburn WFS is thus not considered a freely draining material.

Total Stress and Pore Pressure Measurements

Two total pressure cells measured the stress imposed on the foundation soils under the centerline (C.L.) and shoulder of the embankment (7 m east of CL). Figure 16 presents vertical total stress versus time for both cells. The final difference in stress for the pressure cells (36 kPa) may be due to the effects of soil arching over the earth pressure cell and/or the different location of the cell within the embankment. Using the average measured total stress from both cells, the final height of WFS fill, and average moisture content of 12.6 %, a dry unit weight of 17.4 kN/m³

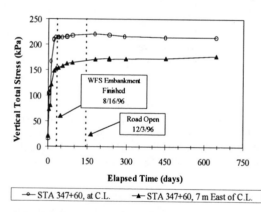

is calculated for the overlying WFS. This estimate agrees with field unit weight measurements presented in Figure 4. Consistent with the total stress measurements, the pore pressure measurements in the subsurface were observed to increase during WFS placement, peak at approximately the same time as the pressure cells, and then return to the baseline readings.

Figure 16: Total Stress vs. Time at the Base of the WFS Embankment

Standard Penetration Testing

Three months after C.R. 206 was opened to traffic, three borings were drilled to evaluate the in situ properties of the WFS, B-borrow, and clay borrow

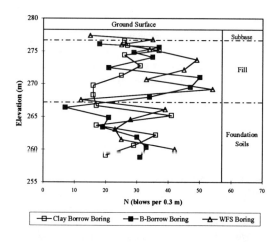

embankments. Figure 17 presents the Standard Penetration Test (SPT) blow counts (N) for each boring as a function of depth below the top of the embankment. Based on penetration resistance, the WFS section had relative densities which were comparable to the B-borrow section. In addition, the WFS had substantially higher blow counts than the clay borrow.

Figure 17: Standard Penetration Test Results

Observations During Construction

Two difficulties encountered during construction were the presence of foreign objects in the WFS (e.g., slag, cores, discarded metal castings, and welding rods) and dust control on the site. The presence of the foreign objects represented a potential for damage to construction equipment tires. To solve this problem, objects were removed at the monofill using a magnetic bar and occasionally by hand at the site. They were also pushed into the fill using the bulldozer blade. By the end of construction, only two punctured tires were reported due to foreign objects. Although foreign objects did not pose a serious problem on this project, we recommend that foundry operators who wish to reuse their WFS for construction should screen such objects.

In general, the WFS arrived on site at a moisture content within the required range for compaction (12-15%). Dust control was not a problem if the WFS was compacted soon after arrival. During the busiest periods of construction, however, some watering of the WFS embankment surface was necessary to limit dust and to reach the required moisture content before compaction. A silt fence was installed to minimize washing of WFS fines off site during heavy rains.

During construction, the WFS material compacted well. No pumping or rutting of the compacted WFS was observed under the wheels of construction equipment, including loaded dump trucks and several large cranes.

OUTLOOK FOR FUTURE WFS REUSE

INDOT saved approximately $145,000 in material costs by using 42,800 m^3 of WFS instead of clay borrow. In addition, the WFS was beneficial in that it exhibited strength properties closer to the B-borrow material, which would have cost about twice as much as the clay borrow for this project. Auburn Foundry donated both the WFS and the trucking costs for the 55 km round trip. Perhaps more significant, Auburn Foundry also accepted long term environmental liability for the WFS embankment. Based on original siting costs and current monofill operating costs, Auburn Foundry estimated that they saved $640,000 by regaining 1.25 years of volume in their monofill. A second estimate, based on much more expensive requirements for siting, building and operating a new monofill, places the savings as high as $1.4 million.

In order for WFS to be used successfully for construction, it must first pass environmental and geotechnical tests to limit the risk of site contamination and poor geotechnical performance. In addition, WFS use will only be economical if: 1) the WFS supplier (i.e., foundry) has enough material of consistent quality for a given project and, 2) the WFS source is within a reasonable distance of the project. The economic benefits of WFS reuse may encourage foundries to assume some or all transport costs to the construction site. In addition, we have recommended that WFSs that pass the necessary environmental and geotechnical quality tests may be used as if they were natural soils, without encasing them between expensive hydraulic barrier layers such as compacted clay, and without special compaction or testing methods (Fox et al. 1997). Regulatory and design decisions with regard to this recommendation will likely play a large role in determining if widespread WFS reuse is to be economically feasible.

CONCLUSIONS

This paper presented laboratory and field test data describing the geotechnical properties of weathered and fresh waste foundry sand (WFS) and the geotechnical performance of a highway embankment constructed using weathered WFS. The WFS used for this project came from Auburn Foundry, Inc. of Auburn, Indiana. The following conclusions were reached as a result of this study:

1. The geotechnical performance of the WFS section was comparable to that of the clean natural sand section, with small internal deformations and a high standard penetration resistance.

2. Dry unit weights of weathered WFS obtained in the field were in general agreement with standard and modified Proctor compaction curves using Method B. Limited test data using Method A procedures did not agree as well with field measurements.

3. Moisture contents obtained using the Speedy Moisture Meter compared well to those obtained using oven-dry samples. Dry unit weights obtained using the

nuclear density gauge showed small errors (≤3%) when compared to those obtained using the sand cone method. This can probably be attributed to differences in measured moisture contents for the nuclear density gauge and Speedy Moisture Meter.

4. As a result of weathering, lower fines contents were measured in WFS samples taken from the top of the monofill. The amount of fines in WFS may affect the strength, hydraulic conductivity, and compaction characteristics of the material.

5. Laboratory and field tests indicated the hydraulic conductivity for compacted WFS ranged from 1×10^{-8} to 7×10^{-7} m/s. In this study, compacted weathered WFS was not considered a free-draining material.

6. WFS dust was controlled during construction by frequent watering of the working surface. Possible damage to construction equipment due to foreign objects in the WFS was initially a concern for this project. However, damage was minimal once the contractor became familiar with the material.

7. Significant potential savings are possible through the reuse of WFS for highway embankment construction. In addition, WFS reuse may help to conserve other natural resources and decrease land acquisitions typically required for project borrow sites.

8. The laboratory and field data presented in this paper are applicable specifically to WFS obtained from a single stockpile. WFS properties will vary depending on the foundry product, casting and recycling processes, storage method, and field construction methods. Appropriate testing is needed for each proposed WFS material. In addition, environmental testing must be performed in order to meet regulatory requirements and minimize the risk of environmental contamination.

ACKNOWLEDGEMENTS

The authors would like to thank a few of the many persons who assisted them during this project: Christopher Burke, Mobil Technology Co.; Eric Triplett and Shu-Hong Chen, Purdue University; Dan Hollenbeck, Auburn Foundry, Inc.; the INDOT field personnel, Fort Wayne, Indiana; and Fox Contractors Corp., Fort Wayne, Indiana. This research project was funded by the Indiana Department of Transportation (INDOT) and the Indiana Cast Metals Association (INCMA).

REFERENCES

ASTM. (1997). *Annual Book of ASTM Standards.* Volume 04.08, American Society for Testing and Materials, West Conshohocken, Pennsylvania.

Fox, P.J., Mast, D.G., Partridge, B.K., and Alleman, J.E. (1997). "Salvaged sand," *Civil Engineering*, Vol. 67, No. 11, 53-55.

INCMA (1992). Indiana Cast Metals Association, Foundry Survey of 1991.

Javed, S. and Lovell, C.W. (1994). "Use of waste foundry sand in highway construction". Final Report JHRP/INDOT/FHWA-94/2J, School of Civil Engineering, Purdue University, West Lafayette, Indiana.

Javed, S., Lovell, C.W., and Hollenbeck, D. (1994). "Spent foundry sand and its uses in Civil Engineering," *Proceedings, Third International Conference on Environmental Issues and Waste Management in Energy and Mineral Production.* Brodie-Hall Research & Consultancy Center Pty. Ltd. Perth, Western Australia: Cuirtin University of Technology, 541-556.

Lovejoy, M.A., Ham, R.K., Traeger, P.A., Wellander, D., Hippe, J., and Boyle, W.C. (1996). "Evaluation of selected foundry wastes for use in highway construction". *Proceedings, 19th International Madison Waste Conference,* Madison, Wisconsin, 19-31.

Mast, D.G. (1997). "Field demonstration of a highway embankment using waste foundry sand." Master of Science in Engineering Thesis, School of Civil Engineering, Purdue University.

Partridge, B.K., Alleman, J.E., Fox, P.J., and Mast, D.G. (1998). "Performance evaluation of highway embankment constructed using waste foundry sand," *Transportation Research Record,* in press.

Trautwein, S. J. and Boutwell, G.P. (1994). "In situ hydraulic conductivity tests for compacted soil liners and caps," *Hydraulic Conductivity and Waste Contaminant Transport in Soil. ASTM STP 1142.* D. Daniel and S. Trautwein, Eds., American Society for Testing and Materials, West Conshohocken, 184-223.

USING WASTE FOUNDRY SAND FOR HYDRAULIC BARRIERS

Tarek Abichou[1], Craig H. Benson[2], Tuncer B. Edil[3], and Brian W. Freber[4]
Members ASCE

Abstract: Green sand is a mixture of uniformly graded fine silica sand, bentonite, and organic binders that is used to make molds for castings in gray-iron foundries. Large quantities of waste green sand in the United States are landfilled each year. Because green sand is a sand-bentonite mixture, there is potential for its use in constructing hydraulic barrier layers used in landfill caps. This paper describes a testing program that was conducted to assess the use of green sand from a central Wisconsin foundry as a barrier material. Specimens of the sand were compacted in the laboratory at a variety of water contents and compactive efforts and then permeated in flexible-wall permeameters. Additional tests were conducted to assess how hydraulic conductivity of the compacted sand is affected by environmental stresses such as desiccation and freeze-thaw. Results of the tests were then compared to hydrologic data collected from two final cover test sections constructed with the green sand.

The hydraulic conductivity of the green sand is sensitive to the same variables that affect hydraulic conductivity of compacted clays (i.e., compaction water content and compactive effort). However, unlike clays, hydraulic conductivities $< 10^{-7}$ cm/sec can be obtained using a broad range of water contents and compactive efforts, including water contents dry of optimum at lower effort. In addition, the hydraulic conductivity of the sand appears unaffected by freeze-thaw or desiccation. Hydrologic data obtained from the final cover test sections show similar results. Very low percolation rates were obtained, even after exposure to harsh Wisconsin winters and summers. In addition, the barrier layers have hydraulic conductivity similar to that measured in the laboratory.

INTRODUCTION

Wisconsin foundries generate about 800,000,000 kg of by-products per year, most of which are landfilled. The high cost of landfilling and the potential uses of foundry by-products have prompted research into their beneficial reuse (Abichou et al. 1998a,b). Gray-iron foundries discard large volumes of waste green sand, which is primarily a mixture of sand and sodium bentonite. Thus, waste green sand has inherent potential for use in construction of hydraulic barrier layers in applications where leachate emanating from the sand will ultimately be collected (e.g., in landfill covers).

[1] Ph.D. Candidate, Dept. of Civil and Environ. Engr., University of Wisconsin-Madison, Madison, WI 53706, Ph. (608)262-6281, abichou@students.wisc.edu

[2] Assoc. Prof., Dept. of Civil and Environ. Engr., University of Wisconsin-Madison, Madison, WI 53706, Ph. (608)262-7242, chbenson@facstaff.wisc.edu

[3] Prof., Dept. of Civil and Environ. Engr., University of Wisconsin-Madison, Madison, WI 53706, Ph. (608)262-3225, edil@engr.wisc.edu

[4] Project Engr., Vierbicher Associates, Inc., 6200 Mineral Point Road, Madison, WI 53705, Ph. (608)233-4131, bfre@vierbicher.com

The objective of this study was to assess the use of waste green sand from a central Wisconsin foundry as a hydraulic barrier material. Specimens of the sand were compacted in the laboratory at a variety of water contents and compactive efforts and then permeated to measure their hydraulic conductivity. Results of these tests were used to define compaction conditions that yield hydraulic conductivity $< 10^{-7}$ cm/sec, the maximum hydraulic conductivity permitted for barrier layers in Wisconsin.

Additional tests were conducted to assess how hydraulic conductivity of the green sand is affected by environmental stresses such as desiccation and freeze-thaw. Results of the tests were then compared to hydrologic data collected by Freber (1996) from two final cover test sections constructed by Vierbicher associates (1996) using the same green sand.

BACKGROUND

Waste Foundry Sand

Foundries use sand in two ways: for molds that form the outside of the casting and in cores that form the internal shapes and cavities within the casting (Javed and Lovell 1994). The sand is bonded by natural clays (e.g., bentonite) or with chemical agents, such as phenolic urethane. Carbon additives, such as sea coal, are also added to control the gas permeability, strength, and other properties of the mix (Barlower 1988, Mackay and Emery 1993). When natural clays bond the sand, it is referred to as green sand. The term "green sand" is used because molten metal is poured into the mold when the sand is damp or "green" (Javed et al. 1994).

A flow chart for a typical green sand molding system is shown in Fig. 1. After the molten metal cools, the casting is removed from the green sand mold and shaken to break out cores and to remove any green sand stuck to the casting. Most of the sand is recycled back into the system (Fig. 1). Core sand is usually crushed and recirculated in the system, which affects the properties of the green sand. As a result, base sand, bentonite, and additives are introduced to the system to maintain the desired properties of the green sand. This results in a gradual accumulation of sand, a portion of which is discarded when the storage capacity of the foundry is reached.

The sand by-product is composed of excess return sand, waste system sand, waste return sand, and core and mold butts. While the by-product contains a variety of constituents, it is often referred to simply as "green sand," which is the primary constituent. Waste green sand usually contains small quantities of contaminants of concern that leach at very low levels (Ham et al. 1981). Guidelines for assessing leachability in Wisconsin can be found in Wisconsin Administrative Code NR Section 538.

Sand-Bentonite Mixtures

Since green sand is primarily sand and bentonite, its hydraulic properties should be similar to those of sand-bentonite (SB) mixtures used in geotechnical practice. Sand-bentonite mixtures are obtained by mixing two very different soils. Fine bentonite particles fill the large voids in the sand plugging the primary flow paths. As more bentonite is added, more of the flow paths become filled. As a result, the hydraulic conductivity generally decreases as the bentonite content increases. The hydraulic conductivity of SB mixtures reaches a minimum when the bentonite fills all voids between the sand particles (Chapuis 1990, Reschke and Haug 1991, Kenney et al. 1992, Howell et al. 1997). Hydraulic conductivity of foundry sand reaches a minimum when bentonite fills approximately 40% of the voids (abichou 1998c)

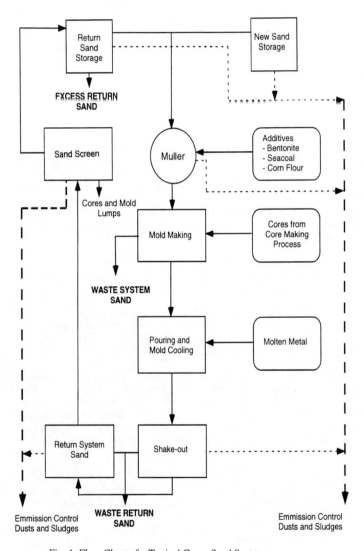

Fig. 1. Flow Chart of a Typical Green Sand System.

Several investigators have studied how water content and compactive effort affect the hydraulic conductivity of SB mixtures. Noir and Wong (1992) indicate that molding water content is not a critical factor in design and construction of sand-bentonite barrier layers because water is not required to break down clods and facilitate remolding, as required for clayey soils (Benson and Daniel 1990). Kraus et al. (1997) report similar findings, and also suggest that the hydraulic conductivity of SB mixtures is not particularly sensitive to compactive effort. The insensitivity to compactive effort is consistent with the lack of correlation between porosity and hydraulic conductivity of SB mixtures reported by Chapuis (1990). However, at low water contents, Kenney et al. (1992) explain that molding water content can affect the hydraulic conductivity of SB mixtures because it affects the distribution of bentonite.

Wong and Haug (1991) and Kraus et al. (1997) investigated how freeze-thaw cycling affects the hydraulic conductivity of SB mixtures. Wong and Haug (1991) report that freeze-thaw cycling causes a decrease in hydraulic conductivity of sand-bentonite mixtures, and that the decrease in hydraulic conductivity is greater for sand-bentonite mixtures with lower bentonite content. They postulate that freeze-thaw cycling helps hydrate the bentonite, and that thawing re-distributes bentonite into spaces between sand grains. Kraus et al. (1997) conducted laboratory and field studies to investigate how freeze-thaw cycling affects the hydraulic conductivity of SB mixtures. In the laboratory and field, they found that the hydraulic conductivity of a SB mixture was unaffected by freeze-thaw action. Kraus et al. (1997) show that ice segregation does not occur in SB mixtures, which prevents cracking and subsequent increases in hydraulic conductivity.

Information on the effect of desiccation on the hydraulic conductivity of SB mixtures is limited. Albrecht (1996) prepared SB mixtures using a well-graded waste rock from a mine and 10% sodium bentonite. After three desiccation cycles, the hydraulic conductivity of the SB mixture was essentially unchanged, suggesting that SB mixtures can be resistant to damage caused by wetting and drying.

Field Demonstration by Freber (1996)

Vierbicher Associates (1996) constructed test sections using the same green sand tested in this study to demonstrate the field performance of barrier layers constructed using green sand. The study included construction of two sets of five 10 m x 10 m test sections underlain by lysimeters. Three of the test sections in each set were constructed with compacted clay barrier layers, whereas the other two test sections contained barrier layers constructed with green sand. The two sets were essentially identical, except for the vegetation. The vegetation was varied to study how it affects the performance of the test sections.

Configurations of the test sections are summarized in Table 1. All test sections were overlain with a 150-mm-thick layer of top soil and seeded. Green sand was used as the barrier layer in Sections D and E and as a barrier protection layer in Sections C, D, and E.

A typical landfill construction quality assurance (CQA) program was followed during construction of the test sections (Freber 1996). The clay barrier layers were compacted to ≥ 90% of modified Proctor maximum dry unit weight, whereas the foundry sand barrier layers were compacted to ≥ 95% (Section D, referred to here as "dense") or 90% (Section E, referred to here as "loose") of maximum dry unit weight. In accordance with the existing specifications of the Wisconsin Department of Natural Resources (WDNR), no water content requirements were used for construction of the barrier layers. All rooting zone layers were compacted to ≥ 80% of standard Proctor maximum dry unit weight. The compaction data collected during construction for Sections D and E are shown in Fig. 3, along with compaction curves for modified, standard, and reduced Proctor efforts.

Two specimens were collected from each barrier layer in Section D (dense) and Section E (loose) using thin-walled sampling tubes (Freber 1996). Hydraulic conductivity of these specimens was determined using ASTM D 5084. Results of these tests are reported by Freber (1996), and are reviewed subsequently in this paper.

Table 1. Configuration of Test Sections (after Vierbicher and Associates, 1996).

Test Section	Description	Function
A	150 mm top soil	Vegetative layer
	600 mm compacted clay	Barrier layer
B	150 mm top soil	Vegetative layer
	900 mm clayey fill	Rooting zone/protection layer
	600 mm compacted clay	Barrier layer
C	150 mm top soil	Vegetative layer
	900 mm green sand	Rooting zone/protection layer
	600 mm compacted clay	Barrier layer
D	150 mm top soil	Vegetative layer
	900 mm green sand	Rooting zone/protection layer
	900 mm compacted green sand	Barrier layer
E	150 mm top soil	Vegetative layer
	2400 mm green sand	Rooting zone/protection layer
	1500 mm compacted green sand	Barrier layer

MATERIALS AND METHODS

Index Properties

Liquid and plastic limits were measured in accordance with ASTM D 4318, except for the time of hydration. ASTM D 4318 specifies that a hydration time of 16 hours is sufficient to conduct Atterberg limits tests on clayey soils. However, testing showed that a week of hydration is required to reactivate "dead" bentonite (bentonite affected by heat during casting) in foundry sands (Kleven 1998). Thus, a one-week hydration period was used for all tests. Particle size distribution was determined using mechanical and hydrometer analyses in accordance with ASTM D 422 after soaking the sand in water for one week. Methylene blue titration (ASTM C 837) was used to determine the bentonite content.

The liquid limit (LL) and plasticity index (PI) of the green sand are 27 and 8. The green sand consists primarily of fine uniform fine sand, but has 12.1% fines (< No. 200 sieve) and a bentonite content (BC) of 8.5%. The particle size distribution is shown in Fig. 2. The green sand classifies as clayey sand (SC) in the Unified Soil Classification System. These index properties suggest that the green sand is suitable as barrier material

(LL $\geq$ 20, BC > 6%) based on correlations between hydraulic conductivity and index properties developed by Abichou et al. (1998c).

Compaction and Hydraulic Conductivity Testing

The green sand was delivered at very low water content (< 2%). Prior to hydration, large objects such as metal pieces and popcorn slag were removed. The material was then spread in a large pan and sprayed with tap water using a spray bottle to achieve the target water content. The moist green sand was then placed in a sealed plastic bag for one week to hydrate.

Specimens of the green sand were compacted using standard (ASTM D 698), modified (ASTM D 1557), and reduced Proctor efforts. Reduced Proctor is the same as standard Proctor, but 15 blows are applied per layer instead of 25 (Daniel and Benson 1990). The compacted specimens were sealed in plastic and left overnight to equilibrate. Then they were placed in flexible-wall permeameters for falling-head hydraulic conductivity testing in accordance with ASTM D 5084. The hydraulic gradients were between 14 and 16 and the average effective stress was 9 kPa. Madison tap water was used as the permeant.

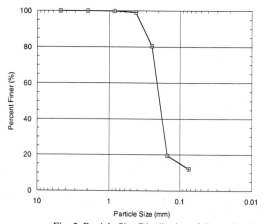

Fig. 2. Particle Size Distribution of Green Sand.

Experience with the green sand indicated that the termination procedures in ASTM D 5084 needed to be modified slightly to ensure equilibrium was obtained. Tests were continued until the last four hydraulic conductivity values were within 25%, inflow equaled outflow, and no trend existed in the hydraulic conductivity or outflow/inflow data. In addition, tests were continued for at least 21 days, even if the aforementioned termination criteria were met at an earlier time. Experience showed that tests that appeared in equilibrium before 21 days occasionally were not actually in equilibrium. In contrast, the ASTM D5084 termination criteria were always suitable when permeation lasted longer than 21 days.

Freeze-Thaw Testing

Tests were conducted to assess how freezing and thawing affects the hydraulic conductivity of green sand. Multiple specimens were compacted at optimum water content using standard Proctor effort. The specimens were then frozen and thawed three-dimensionally at zero overburden pressure following the free-standing procedure in accordance with the methods described in ASTM D 6035.

After reaching the desired number of freeze-thaw cycles, the specimens were removed from the freezer and placed in flexible-wall permeameters. Twenty-four hours of thawing were allowed before falling-head hydraulic conductivity tests were initiated. The hydraulic conductivity tests were conducted following ASTM D 5084 using the procedures mentioned previously.

Desiccation Tests

Tests were also performed to assess how wetting and drying affect the hydraulic conductivity of green sand. Specimens were prepared at optimum water content using standard Proctor effort, and then permeated in flexible-wall permeameters in accordance with ASTM D5084 using the method described previously.

After initial permeation, the specimens were removed from the permeameters and left to air-dry in the laboratory. The specimens were weighed daily to monitor moisture loss. Once the weight of a specimen ceased changing, it was placed in a flexible-wall permeameter for hydraulic conductivity testing. This procedure was repeated for five wet-dry cycles.

RESULTS AND ANALYSIS

Compaction and Hydraulic Conductivity

Results of the compaction and hydraulic conductivity tests are summarized in Table 2 and are graphed in Figs. 3 and 4, along with data reported by Freber (1996). Hydraulic conductivity of the laboratory-compacted specimens varies less than one order of magnitude as the compaction water content is varied, which is a much smaller range than that of clayey soils (Benson and Daniel 1990). In addition, all three compactive efforts resulted in similar hydraulic conductivities, ranging from 2.3×10^{-9} to 2.5×10^{-8} cm/sec. Thus, compactive effort had less effect on hydraulic conductivity of green sand than it typically has on the hydraulic conductivity of clayey soils. This behavior is consistent with the behavior of sand-bentonite mixtures reported by Wong and Haug (1992) and Kraus et al (1997). Moreover, the insensitivity to compaction water content and effort suggests that barrier layers constructed with green sand requires less water addition and moderate compactive effort.

Similar hydraulic conductivities are reported by Freber (1996) for the specimens from the test sections. Specimens collected from Test Section D had hydraulic conductities of 8.3×10^{-9} and 1.3×10^{-8} cm/sec and those from Section E were 5.4×10^{-8} and 6.8×10^{-8} cm/sec (Freber 1996). Hydraulic conductivities of the specimens from Section D (dense) are in the same range as the hydraulic conductivities of the laboratory-compacted specimens. Hydraulic conductivities of the specimens from Section E (loose) are slightly higher, probably because the green sand was placed with less effort.

Field hydraulic conductivities were estimated from the maximum percolation rates reported by Freber (1996) and Vierbicher Associates (1996) using a unit hydraulic gradient. The maximum percolation rate was used because it most likely corresponds to nearly saturated conditions. The field hydraulic conductivity of Section D varies from 1.3×10^{-8} to 1.5×10^{-8} cm/sec, whereas the field hydraulic conductivity of Section E varies from 3.3×10^{-8} to 6.1×10^{-8} cm/sec. The field hydraulic conductivities of Section D (dense) are in the same range as the hydraulic conductivities of the laboratory-compacted specimens. The field hydraulic conductivities of Section E (loose) are higher. These hydraulic conductivities are similar to those obtained by Freber (1996) from the thin-wall sampling tubes, which suggests that thin-walled sampling tubes can be used for quality control testing of barrier layers constructed with green sand.

Table 2. Molding Water Contents, Dry Unit Weights, and Hydraulic Conductivities of Green Sand specimens.

Specimen	Molding Water Content (%)	Dry Unit Weight (KN/m³)	Hydraulic Conductivity (cm/sec)	Compaction	Notes
1 M	7.2	18.2	1.6×10^{-8}	Modified Proctor	Lab Compacted
2 M	8.7	18.1	1.3×10^{-8}	Modified Proctor	Lab Compacted
3 M	10.9	18.1	1.4×10^{-8}	Modified Proctor	Lab Compacted
4 M	12.5	18.1	2.3×10^{-9}	Modified Proctor	Lab Compacted
5M	13.9	17.4	2.4×10^{-9}	Modified Proctor	Lab Compacted
1 S	10.5	17.3	1.3×10^{-8}	Standard Proctor	Lab Compacted
2 S	12.6	17.1	8.1×10^{-9}	Standard Proctor	Lab Compacted
3 S	14.2	17.1	6.0×10^{-9}	Standard Proctor	Lab Compacted
4 S	16.1	16.7	5.0×10^{-9}	Standard Proctor	Lab Compacted
5 S	18.5	16.9	3.9×10^{-9}	Standard Proctor	Lab Compacted
1 R	12.0	16.5	2.5×10^{-8}	Reduced Proctor	Lab Compacted
2 R	14.2	16.7	1.1×10^{-8}	Reduced Proctor	Lab Compacted
3 R	16.2	16.5	3.9×10^{-9}	Reduced Proctor	Lab Compacted
4 R	18.5	16.3	4.1×10^{-9}	Reduced Proctor	Lab Compacted
5 R	19.6	16.2	2.6×10^{-9}	Reduced Proctor	Lab Compacted
6 R	21.9	N/A	3.1×10^{-9}	Reduced Proctor	Lab Compacted
Section 1 D	15.4	17.9	8.3×10^{-9}	Dense	Field Compacted
Section 2 D	10.7	17.9	1.3×10^{-8}	Dense	Field Compacted
Section 1 E	25.8	15.6	6.8×10^{-8}	Loose	Field Compacted
Section 2 E	13.3	17.4	5.4×10^{-8}	Loose	Field Compacted

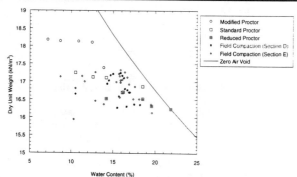

Fig. 3. Compaction curves and field compaction data for Sections D and Sections. E

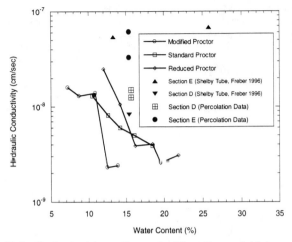

Fig. 4. Hydraulic Conductivity vs. Compaction Water Content for Laboratory-Compacted Specimens, Undisturbed Specimens Removed from the Barrier Layer Constructed with Green Sand, and Percolation Data.

Freeze-Thaw

Results of the freeze-thaw tests are reported in terms of hydraulic conductivity ratio (K_r), which is defined as the ratio of the hydraulic conductivity (K_n) after n freeze-thaw cycles to the initial hydraulic conductivity (before freeze-thaw cycling). Hydraulic conductivity ratio (K_r) versus number of freeze-thaw cycles is shown in Fig. 5, along with SB data from Kraus et al. (1997) and Wong and Haug (1992). The hydraulic conductivity of the green sand appears to be unaffected by freeze-thaw. The ratio K_r ranges from 0.7 to 1.2.

Although thermal conditions were not measured in the test sections, the test sections have probably frozen because the frost depth near Reedsburg, Wisconsin, typically varies from 1.5 to 2 m. Vierbicher Associates (1996) report that the barrier layers constructed with the green sand had no cracks or fractures after exposure to Wisconsin winter. Hence, freeze-thaw appeared to have no detrimental impact on the barrier layers constructed in the field. The low hydraulic conductivities observed in the field are consistent with this observation (Fig. 4).

Desiccation

Results of the desiccation tests are also described in terms of hydraulic conductivity ratio (K_r) following the definition used in the freeze-thaw tests. Hydraulic conductivity ratio (K_r) vs. number of wet-dry cycles is shown in Fig. 6, along with the SB data from Albrecht (1996). The ratio K_r for the foundry sand ranges from 0.45 to 3.5 within five wet-dry cycles, which suggests that the foundry sand is more resistant than clayey soils to deterioration associated with wet-dry cycles. For example, Albrecht (1996) shows that the hydraulic conductivity of compacted clays increases 100 to 1000 times when subjected to wet-dry cycling.

Freber (1996) reports percolation rates from the test sections containing green sand from 1992 to 1996. Little change in percolation rate occured from year to year, suggesting that weather had not affected the test sections. He also reports that the barrier layers constructed with green sand contained no cracks or fractures after exposure to Wisconsin summers. In contrast, the compacted clay barriers in the test sections were severely cracked by desiccation (Albrecht 1996).

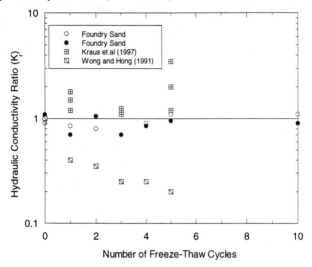

Fig. 5. Hydraulic Conductivity Ratio vs. Number of Freeze-Thaw Cycles.

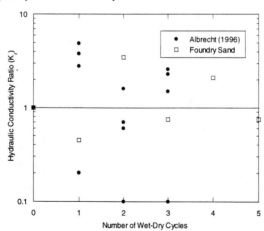

Fig. 6. Hydraulic Conductivity Ratio vs. Number of Wet-Dry Cycles.

Percolation from Test Sections

Vierbicher Associates (1996) have monitored percolation emanating from the test sections since construction (1991). The average annual percolation rate through March 1996 is shown in Fig. 7. These percolation rates are the average for the two different vegetation schemes. Vegetation had no distinct effect on percolation rate.

The percolation rates for the test sections constructed with green sand (C, D, and E) are approximately two orders of magnitude lower than those for the test sections constructed with compacted clay (A, B), regardless of whether the green sand was used in both the barrier and protection layers (D, E), or solely in the protection layer (C) (Fig. 7). In addition, percolation rates for the test sections with the green sand barrier layers are similar, regardless of the thickness of the barrier layer or thickness of the protection layer.

Cracks in the clay barrier layers were apparently responsible for the high percolation rates measured in Sections A and B. Mineral deposits were found on the smooth fracture surfaces. The fractures also had a moist surface, but the intact clay between the fractures was dry. In addition, roots were found within the fractures, but not in the intact clay between the fractures. Large (30-cm-diameter) block specimens were removed from the clay barrier in Section A to assess the field hydraulic conductivity. The hydraulic conductivity of the clay was found to be 5×10^{-5} cm/s, on average, which is more than two orders of magnitude higher than the as-compacted hydraulic conductivity (Albrecht 1996).

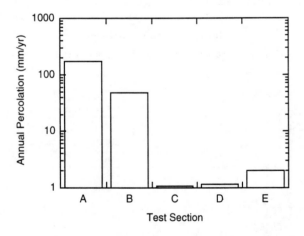

Fig. 7. Average Annual Percolation from Test Sections.

SUMMARY

Gray-iron foundries landfill large quantities of waste green sand each year. Waste green sand is a by-product of casting, and is primarily a mixture of fine sand and bentonite. This study evaluated the beneficial use of a green sand from a central Wisconsin foundry for construction of hydraulic barrier layers used in landfill final covers. Index properties, compaction tests, and hydraulic conductivity tests were conducted on the green sand. Tests were also conducted to assess the resistance to damage caused by freeze-thaw cycling and desiccation. Results of these tests were then compared to field data collected from final cover test sections constructed with the green sand by Vierbicher Associates as by reported by Freber (1996).

Laboratory testing showed that the green sand has a liquid limit and a plasticity index conducive to achieving the low hydraulic conductivity necessary for construction of barrier layers. Hydraulic conductivity testing of laboratory-compacted specimens showed that the green sand can be compacted to very low hydraulic conductivity (10^{-9} to 10^{-8} cm/sec) using a broad range of compactive efforts and a wide range of water contents. In addition, freeze-thaw and desiccation testing suggest that the hydraulic conductivity of the green sand is unlikely to be affected by frost or desiccation.

Result of the field tests conducted by Freber (1996) show that final covers constructed with green sand barrier layers perform better than those constructed with compacted clays. Percolation rates from the test sections constructed with green sand were significantly lower than the percolation rates from test sections constructed with compacted clay even after exposure to several harsh winters and summers, during which frost action and desiccation certainly occurred. Even after exposure, field hydraulic conductivities of the green sand barrier layers were similar to those measured in the laboratory on laboratory-compacted specimens and on undisturbed specimens from the field. In summary, green sands offer a superior barrier than conventional clays and possibly at lower cost.

ACKNOWLEGMENT

Financial support for the study described in this paper was provided by the University of Wisconsin System Solid Waste Research Program (SWRP). Grede Foundries (Reedsburg, Wisconsin) provided samples of their waste sand. They also conducted the methylene blue tests to determine the bentonite content of the foundry sand. However, the findings described in this paper are solely those of the authors. Endorsement by SWRP or Grede Foundries is not implied.

REFERENCES

Abichou, T., Benson, C., and Edil, T. (1998a), "Database on Beneficial Reuse of Foundry By-Products." Environmental Geotechnics Report 98-3, Dept. of Civil and Environmental Engineering, University of Wisconsin-Madison.

Abichou, T., Benson, C., and Edil, T. (1998b), "Database on Beneficial Reuse of Foundry By-Products." In this GSP on Recycled Materials in Geotechnical Applications, GeoCongress-98, Boston.

Abichou, T., Benson, C., and Edil, T. (1998c), "Beneficial Reuse of Foundry Sands in Construction of Hydraulic Barrier Layers." Environmental Geotechnics Report 98-2, Dept. of Civil and Environmental Engineering, University of Wisconsin-Madison.

Albrecht, B. A. (1996), "Effect of Desiccation on Compacted Clays", M. Sc. Thesis, University of Wisconsin-Madison. Madison, WI.

Benson, C. H. and Daniel, D. E. (1990), "Influence of Clods on Hydraulic Conductivity of Compacted Clay," J. Geotechnical Engineering, ASCE, Vol. 116, No. 8, pp. 1231-1248.

Bralower, P. M. (1988), "The Use and Misuse of Green Sand Additives, Part 2," Modern Casting, Vol. 78, No. 10, pp. 46-48.

Chapuis, R. P. (1990), " Sand-Bentonite Liners: Predicting Permeability from Laboratory Tests," Canadian Geotechnical Journal, Vol. 27, No. 1, pp. 47-57.

Freber, B. W. (1996), "Beneficial Reuse of Selected Foundry Waste Material," Proceeding of 19th International Madison Waste Conference, Madison, WI, No. 13, Sept. 1996, pp. 246-257.

Howell, J. L., and Shackelford, C. D. (1997), "Hydraulic Conductivity of Sand Admixed with Processed Clay Mixtures," Proceedings, Fourteenth International Conference on Soil Mechanics and Foundation Engineering, Hamburg, Sept. 6-12, 1997, Balkema, Rotterdam, Vol. 1, pp. 307-310.

Javed, S. and Lovell, C. W. (1994), "Use of Waste Foundry Sand in Highway Construction," Report JHRP/INDOT/FHWA-94/2J Final Report. Purdue School of Engineering, West Lafayette, Indiana.

Kenney, T. C., Van Veen, W. A., Swallow, M. A., and Sungaila, M. A. (1992), "Hydraulic Conductivity of Sand-Bentonite Mixtures," Canadian Geotechnical Journal, Vol. 29, No. 3, pp. 364-374.

Kleven, J. R. (1998), "Mechanical Properties of Excess Foundry System Sand and Evaluation of its Use in Roadway Structural Fill." M. Sc. Thesis, University of Wisconsin-Madison. Madison, WI.

Kraus, J. F., Benson, C.H., Erickson, A. E., and Chamberlain, E. J. (1997), "Freeze-thaw Cycling and the Hydraulic Conductivity of Bentonitic Barriers," J. Geotechnical and Geoenvironmental Engineering, ASCE, Vol. 123, No. 3, pp. 229-238.

Mackay, M., and Emery, J. J. (1993), "Use Of Waste, Surplus Materials and Byproducts in Transportation Construction," Symposium Proceedings Recovery and Effective

Reuse of Discarded Materials and By-products for Construction of Highway Facilities, pp. 1.17-1.28.

Vierbicher Associates (1996), "Final Report: Beneficial Reuse of Selected Foundry Waste Material," Prepared for Wisconsin Dept. of Natural Resourses, March 1996.

Wong, L. C., and Haug, M. D. (1991), "Cyclical Closed-System Freeze-Thaw Permeability Testing of Soil Liner and Cover Materials," Canadian Geotechnical Journal, Vol. 28, pp. 784-793.

Recycled Materials for Embankment Construction

C. Vipulanandan[1] M.ASCE and M. Basheer[2]

Abstract

Obtaining good quality soils at reasonable costs for use in embankment construction is becoming difficult and hence, attention is directed towards finding alternative materials. On the other hand, non-hazardous solid wastes are being generated in large quantities in different forms around the nation. Increasing disposal cost, constricting landfill space and stringent environmental regulations call for exploring large volume utilization alternatives. Hence, utilization of these solid waste materials in embankment construction is a promising alternative to explore in the interests of highway and environmental agencies. Several potential solid waste materials (tire chips, fly ash, bottom ash, scrubber base, flourogypsum) were selected for this investigation. Their engineering and environmental characteristics were determined using geotechnical tests and two leaching tests including the TCLP test recommended by the USEPA. Tire chips and fly ash (C) were mixed with sand and a clayey sand and some of their properties were determined. Based on the CBR test results most of the materials tested were suitable for subgrade construction except for fly ash (F) and scrubber base (F) where the CBR values were close to 1, under soaked condition. All the solid waste materials studied are environmentally benign based on the level of metal leached. Sulfate concentration in leachates were higher than 1000 ppm for flourogypsum and scrubber bases.

Introduction

Non-hazardous solid waste in the amount of 4.6 billion tons is being generated each year in different forms around the nation (NCHRP, 1994a). Disposal of this solid waste is a major problem due to shrinking landfill space, environmental hazard and rising disposal costs. Selective utilization of these solid waste in highway embankment construction can serve the dual purpose of ameliorating the disposal problem to a certain extent and can be a source of alternative construction material for highways in view of the diminishing supplies

Professor and Director[1] and Graduate Student[2], Center for Innovative Grouting, Materials and Technology (CIGMAT), Department of Civil and Environmental Engineering, University of Houston, Houston, Texas 77204-4791. Phone:(713)743-42678. email: cvipulanandan@uh.edu

of natural high quality soils at reasonable cost [Bloomquist et al. 1993].

Embankment construction utilizes enormous quantities of material and the cost of construction has been consistently rising due to depletion of good borrow material. Hence, if an alternative material has to be considered then it should be available easily in large quantities and at a low cost. Non-hazardous waste materials such as coal combustion by-products, flourogypsum and scrap tires comprise about 25% of the non-hazardous waste currently being generated in the U.S [NCHRP, 1994a]. These materials are either landfilled or stockpiled and are available readily at a low cost.

In this study an extensive literature review was performed to identify the potential recycled materials for embankments, different embankment configurations and construction methods [MPCA, 1990; Bosscher et al. 1992; Ahmed, 1993; Maryland DOT, 1993; Hoppe, 1994; Vipulanandan et al. 1996]. In using waste materials for embankment constructions, normal construction techniques may need to be altered depending upon the performance of the material. Since high degree of variability is inherent in these materials, an extensive testing program was required to evaluate the waste materials for near future applications. The geotechnical characterization included the determination of shear strength parameters, compaction characteristics, CBR, permeability, grain size and other physical parameters of solid waste materials and waste-soil mixtures. The leaching studies were performed to identify the potential contaminants.

Literature Review

As part of the literature survey, case studies were performed to determine the widely used waste materials in embankments, different embankment configurations and construction techniques adopted in the past. A total of 39 case histories from 16 states and Europe, pertaining to the use of waste materials in the highway embankments, have been documented. These case histories were obtained from journals, State Department of Transportation (DOT) records and from certain environmental agencies. Fly ash has been the most widely used material since the early 70's till the late 80's. The use of scrap tires in embankment construction began in the 90's especially due to the scrap tire management programs started by different states. This program encouraged tire shredding into tire chips that are being incinerated for energy recovery, production of crumb rubber for use as binder in pavements and for bulk usage such as embankment construction.

The different types of embankment configurations that have been used are shown in the Figure 1. Type A has been the most frequently used configuration (76% of the case studies) because the waste material is placed in the core with the help of geotextiles [Basheer et al. 1996]. The top soil cover helps to distribute the loads, and the side slope cover helps reduce water infiltration into the embankment core and aids in stabilizing the slope with vegetation. Type B has been used in only 12% of the case studies with different types of alternating recycled materials. Type C (12%) consist of mixture of waste and naturally available soil. Uniform mixture in this kind of construction is difficult and may be a limitation in its use. It has been reported that this configuration with a good control on the homogeneous mixing performs better than Type A configurations using tire chips as the waste material in 50/50 mixture with naturally available soil (Bosscher et al. 1994).

The recycled materials selected for testing in this study was based on availability, cost and engineering aspects. Shredded tires, coal combustion by-products (fly ash, bottom ash and scrubber base) and flourogypsum have been identified as the potential materials. Quantities generated, current applications, percentage recycled and the material characteristics of the solid wastes have been discussed in detail in the following sections.

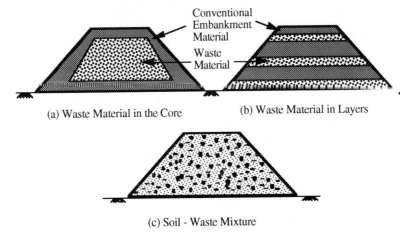

(a) Waste Material in the Core   (b) Waste Material in Layers

(c) Soil - Waste Mixture

Figure 1. Different Embankment Configurations

Tire Chips

About 242 million scrap tires are generated in the United States annually and there are about 2 billion that are stockpiled (NCHRP, 1994b). The current waste disposal practice is that, out of the 242 million tires that are discarded annually, 5 percent are exported, 6 percent recycled, 11 percent incinerated and 78% are landfilled, stockpiled, or illegally dumped (Ahmed & Lovell, 1993). Whole tires are not easily disposed due to their poor compressibility, potential combustibility and associated toxic fumes (Bosscher et al., 1992); hence, in order to improve the ease of disposal, tires are shredded into tire chips. The tire chips are irregularly shaped and may either have reinforcing wire embedded in them completely or may not have any wire at all. The cost of tire chips varies from $5 - $50 per cubic yard depending on the quality of tire chips with the upper limit of cost being for sizes less than 0.25 inch. The specific gravity of tire chips varies from 0.8 g/cc to 1.28 g/cc (Ahmed, 1993). Tire chips having metal wires embedded in them have higher unit weights. The current applications of tire chips is in rubber modified asphalt and in incineration fuel for various incinerators and cement kilns. The 78% of the landfilled and stockpiled tire chips are a menace because they are breeding grounds for mosquitoes and a potential fire hazard. Since they are light in weight and occupy a huge volume, they can be used in embankments with certain construction techniques.

Fly ash

Fly ash is produced during the combustion of ground or powdered coal (ASTM C 618). The chemical compositions and physical appearance vary with the coal properties, burning processes, and collection procedures. In general, fly ash consists of heterogeneous combinations of glassy and crystalline phases. Its principal constituents are silica, alumina, iron oxide, lime and carbon. According to ASTM C 618, fly ash is classified into Class-C and Class-F based on the content of calcium oxide (CaO) and other metal oxides. If the fly ash has a CaO content less than 10%, it is technically considered Class-F; otherwise, Class-C. Class-F fly ash is produced by burning anthracite or bituminous coal while Class-C fly ash is produced by burning lignite or sub-bituminous coal. The principal active component is siliceous or aluminosilicate glass for Class-F fly ash, and calcium aluminosilicate glass for Class-C fly ash (Diamond 1983, Roy, Luke; and Diamond, 1984). Approximately 75 million tons of coal ash is produced in the U.S. annually (NCHRP, 1994a). Current applications of fly ash are in cement, concrete, grout, soil remendment and waste stabilization. Fly ash is a non-plastic, silt sized with a specific gravity varying between 2.4 - 2.7 g/cc (DiGioia and McLaren, 1987). Hence, its use in the highway industry can release the pressure on the landfills and reduce the material cost in embankment projects.

Bottom ash

Millions of tons of bottom ash is produced annually in the same process as the production of fly ash. Exact quantities being recycled is unknown. The cost of bottom ash is in the range of $5 per ton and is available from most of the coal fired power plants. Current applications of bottom ash are in road base construction, construction of drainage layers, anti skid and ice-control purposes. The specific gravity range of bottom ash ranges from 1.9 to 3.4 and is a function of its chemical composition. High carbon content results in low specific gravity and high iron content results in higher specific gravity. The principal constituents of bottom ash are silica (SiO_2), alumina (Al_2O_3), and iron oxide (Fe_2O_3). There are smaller quantities of calcium oxide (CaO), magnesium oxide (MgO), potassium oxide (K_2O), sodium oxide (Na_2O) and sulfur trioxide (SO_3). Bottom ashes have been characterized by EP toxicity test as non hazardous and it has been reported that the salt concentrations of bottom ash extracts have minimal effects on ground water quality (Lovell et al. 1991).

Scrubber base

Scrubber base is the mixture of FGD (Flue gas desulfurization sludge) and fly ash (class F or C) in a 50/50 proportion (Snow, et al. 1987). FGD is the by-product of an electric utility boiler's pollution control system, commonly called a scrubber system. Blending of FGD with fly ash (which converts it to scrubber base) is necessary to dry and stabilize it sufficiently for land filling. It is available at a cost of $7 per ton. Current applications of scrubber base are in roadway base course construction and agricultural soil amendment (Texas Recycle 2, 1994). Environmental concerns pertaining to the use of scrubber base would be similar to that of usage of fly ash.

Flourogypsum

About 40 million tons are produced annually in the U.S. Flourogypsum is a much lighter material than most soils and is relatively well graded with about 75% fines passing # 200 sieve. It contains about 71% gypsum ($CaSO_4. 2H_2O$) mostly in half-hydrate and anhydrite forms (Brown and Associates, 1987).

Testing Program

The experiments were divided into geotechnical tests and environmental (leachability) tests. The properties of interest were specific gravity, particle-size distribution, index properties, compaction characteristics, unconfined compressive strength, cohesion and friction angle.

Geotechnical Tests

Index Properties: Specific gravity, grain size distribution and Atterberg limits were determined in order to characterize the index properties of the materials being used in the test. The results from specific gravity and Atterberg limits tests are summarized in Table 1.

Table 1. Specific Gravity and Atterberg Limits for Solid Waste Materials.

Material	Specific gravity	LL	PL	PI	USCS Classification
Fly ash (C)	2.79	14	NP	-	ML
Fly ash (F)	2.52	26	NP	-	ML
Flourogypsum	2.10	47	39	8	ML
Scrubber base (C)	2.25	27	21	6	SW-ML
Scrubber base (F)	2.49	58	33	25	SP-MH
Bottom ash	2.55	NP	NP	-	SP
Tire Chips	1.28	NP	NP	-	GP

Particle size distributions for the solid waste materials are shown in Fig. 3. Tire chips and bottom ash are uniformly graded whereas scrubber base (C) and flourogypsum are gap graded. Both the fly ashes are well graded with about 85% of the particles in fly ash (C) and 75% of the particles in fly ash (F) finer than No. 200 sieve (0.075 mm). The results indicate that fly ash (C) is the finest with 28% which is followed by cement and fly ash (F) having fineness of 20% and 15% respectively. Rest of the materials have no fines which are below he size of No. 325 sieve. Tire chips and bottom ash are relatively coarse and can be used for filter construction. Based on the index properties and particle size distribution, the recycled materials are classified according to the USCS system. Fly ash (Class C and Class F) are non plastic and have their liquid limits close to their optimum moisture contents. However, based on their particle size and liquid limit they have been classified as ML (low plasticity silts). Bottom ash and tire chips are granular materials and have been classified based on their particle sizes and gradation.

Compaction: Maximum dry density and optimum moisture content was determined for all the waste materials using the standard 4 inch Proctor mold according to the ASTM D 698. Compaction behavior of pure tire chips and their combinations with clayey sand was studied using a 6 in. x 8 in. size mold. The results of the compaction tests are summarized in the Table 2 and Figure 3.

Since moisture density relationship could not be determined for tire chips, bulk density were determined for varying proportions of tire chips in tire chips -sand and tire chips-clayey soil mixtures. It can be seen from Fig. 3 that the variation of bulk density is almost linear with the percentage of tire chips in the mixture. For tire chips-soil mixtures with 30 to 50% tire chips the strains varied between 4 and 10% on loading up to 0. 7 MPa (100 psi).

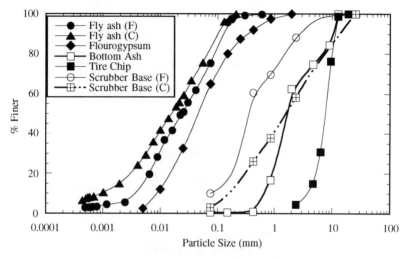

Figure 2. Particle size Distribution of Recycled Materials

Table 2. Standard Compaction Test Results

S. No.	Material	Optimum Moisture (%)	Maximum Dry Density (pcf)
1	Fly ash (C)	12	132
2	Fly ash (F)	10	110
3	Scrubber base (C)	36	76
4	Scrubber base (F)	24	91
5	Bottom ash	26	93
6	Flourogypsum	18.5	92

Shear Strength: Consolidated drained triaxial tests were performed on Harvard miniature samples (1.4" x 2.8") of waste materials at three confining pressures of 5, 10 and 20 psi. Tests were conducted according to ASTM D 2850. Behavior of the tire chips and their mixtures was studied using sample of 150 mm (6-in.) diameter in a large triaxial cell. The effective shear strength parameters for the solid waste materials that were tested are summarized in the Table 3.

Fly ash (F) is non-cohesive and its shear strength is primarily the result of internal friction. In contrast to this behavior, fly ash (C) can develop considerable cohesive shear strength due to cementitious reactions. This cohesion is the dominant shear strength of fly ash (C) (McLaren and DiGioia, 1987). Scrubber base (F) shows an effective friction angle and cohesion which is close to that of

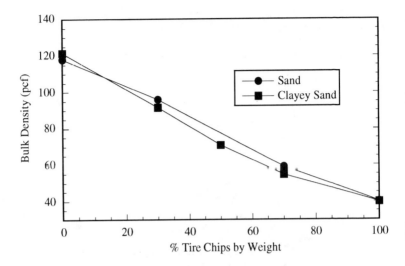

Figure 3. Variation of Bulk Density with Different Proportions of Tire chips in Tire chips - Soil Mixtures

Table 3. Shear Strength Parameters for the Recycled Materials

Materials	Cohesion, c (psi)*	Friction angle (ϕ)	Dry Density (pcf)
Fly ash (C)	28	61°	132
Fly ash (F)	3	39°	110
Scrubber base (F)	7	33°	91
Scrubber base (C)	13	41°	76
Flourogypsum	8	40°	93
Bottom ash	6	45°	92
Tire Chips	0	30°	40
30% Tire chips + 70% Sand	0	41°	96

* 1 psi = 6.9 kPa; 1.6 kN/m^3 = 10 pcf

FGD sludges (Krizek et al. 1987). This means that the addition of fly ash (F) in 50-50 proportion to FGD (basic composition of scrubber base (F)) has little influence on its effective shear strength parameters. In the case of scrubber base (C), (fly ash (C) and FGD in a 50-50 mixture) fly ash (C) was effective in enhancing the shear strength parameters. The value of shearing resistance for bottom ash was higher than those for conventional granular sands. This difference can be attributed to the rough surface texture and angularity of the bottom ash particles that allows for higher degree of interlocking during the shearing process. Tire chips and it's sand mixtures are non-cohesive, the shear strength in these materials is mainly attributed to angle of friction which is higher than most soils.

Permeability: Constant head permeability tests were performed using a 100 mm (4 inch) diameter fixed wall double ring permeameter. A constant hydraulic gradient of 100 was used. This test is very

Table 4. Results of the Permeability Test

Material	Wet* Density (pcf)	pH[1]	Permeability (cm/s)
Fly ash (C)	130	12.2	9×10^{-9}
Fly ash (F)	124	9.9	7×10^{-6}
Scrubber Base (F)	115	8.4	8×10^{-5}
Scrubber Base (C)	104	8.8	1×10^{-5}
Flourogyps um	115	7.7	2×10^{-6}
Bottom ash	82	-	5×10^{-3}

* $1.6 \text{ kN/m}^3 = 10 \text{ pcf}$

similar to the constant head permeability test (ASTM D 2434) except for the sample size. The pH of the effluent fluid was monitored throughout the test. These tests were continued till the coefficient of permeability reached a constant value. The results of the permeability test are summarized in Table 4 with the average value of pH of the effluent. The permeability of bottom ash was determined using a falling head permeameter. The permeability was in the range of granular soils. Among the materials tested, fly ash (C) had the lowest permeability (10^{-9} cm/s) due to the self hardening of fly ash (C) with time. Scrubber base (C) and scrubber base (F) had highest permeability among the finer material that were tested. This is due to the high porosity of the material in compacted form.   The low permeability of solid waste materials, except tire chips and bottom ash lessen the probability of extensive ground water percolation or infiltration from runoff, and the consequent danger of soluble material being leached out of the fill. In addition, low permeability means a high degree of runoff therefore precautions to prevent erosion of side slopes should be taken (Gray and Lin, 1972).

CBR Test: This test was performed on all the waste materials selected except on tire chips. The test was performed in accordance to ASTM D 1883. All the samples were allowed to soak for a period of 96 hours before testing. As shown in Fig. 4, fly ash (F) and scrubber base (F) have the lowest CBR values in soaked condition. This can be attributed to very low cohesion between the individual particles and localized liquefaction near the base of the piston. Fly ash (F) and scrubber base (F) can be rated as "very poor" for use as a base course or a subbase. Fly ash (C) had the highest CBR values with and without soaking. Scrubber base (C) can be rated as "fair" and it has a potential to be used for subgrade construction. Also note that soaking did not affect the CBR value of Scrubber base (C). Bottom ash and flourogypsum fall in the category of good base courses and subbases and had higher CBR values than scrubber base (C).

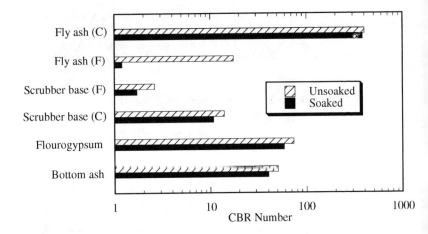

Figure 4. CBR values for the Solid Waste Materials

Environmental Tests

The environmental characteristics were determined by conducting leaching studies both in acidic and neutral leaching solutions. The leaching tests performed in this study are :

Water Leaching Test: Solid waste materials were mixed with distilled water in the ratio of 1:4 by weight in 2 liter bottles. In each bottle, 250 g of solid waste material was mixed with one liter of distilled water. The leaching test was continued for 100 days. At the end of test the leachate was filtered through 0.45 micron filter paper and the filtrate was analyzed for metals and sulfates using ICP (Inductively coupled plasma) and IC (Ion chromatography). Passing the water test, in many states, qualifies the solid waste material as a Class 3 waste material that is non-hazardous.

The leachate was analyzed for six metals (Ba, Cr, Pb, Zn, Cu and Ca). The only detectable metal in almost all the samples was calcium. Barium was detected only in the leachate of cement sample. Rest of the metals were below the detection limit of the ICP which is 0.1 ppm. Three replicates were performed both for the metal analysis and for the detection of sulfate. It can be seen (Fig. 5) that both scrubber bases and flourogypsum have high concentrations of sulfate in their leachates which is expected because of the large quantity of calcium sulfate in these materials. Similarly, the calcium concentration is also relatively high in these materials. The presence of calcium in the leachate of tire chips may be due contamination. For Class 3 waste (in Texas) the Maximum Contaminant Level (MCL) for barium Ba (II), chromium and lead are 1, 0.1 and 0.05 mg/L respectively. There is no MCL for calcium Ca (II).

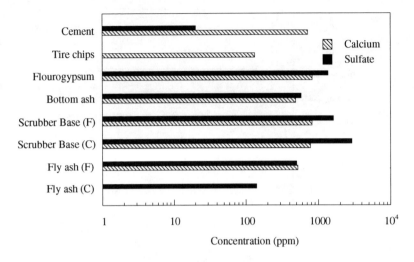

Figure 5. Water Leachate Test Results

TCLP (Toxicity Characteristics Leachate Procedure): Tests were performed on recycled waste materials smaller than 9.5 mm (passing No. 10 sieve). The extraction fluid for the leachate test was selected based on the pH of the mixture of de-ionized (DI) water and recycled material. Each material selected in this study was evaluated separately to select the leaching solution. All recycled material exhibited a pH of 7 or higher. Each TCLP test sample (approximately 100 g) was mixed with DI water in a 2 liter vessel, with a solid to liquid ratio of 1:20 by weight, and then agitated in a rotary tumbler at 30 rpm for 18 hours. The test was continued for a total of seven days. Metal analysis and sulfate determination was performed using the ICP and IC analytical tools for the same samples as in the case of the Water Leaching test. Similar to the water leachate samples, most of the metals that were being investigated were non detectable except calcium and barium. Barium was present only in fly ash (C) and cement with concentrations of 0.2 and 0.35 ppm respectively. These concentrations are much below the TCLP limits for barium which is 100 ppm. Although, the solid to liquid ratio is much lower in this test (1:20) the concentration of calcium for most of the materials is higher in this test than in the case of Water test which is due to the rigorous agitation and acidic leachate solution. Sulfate concentrations are lower in this case compared to the samples leached in water. This indicates that sulfate is not readily leached out from the solid matrix with the same ease as metals in acidic environment. The concentration of calcium leached out from cement is much higher than that in the Water leachate samples and very high compared to the other samples that were tested. Calcium was present in the leachate of fly ash (C), indicating that bound calcium could be leached out in acidic environment unlike the alkaline environment in case of Water test.

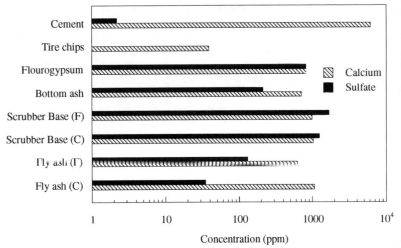

Figure 6. TCLP Test Results

Recycled Material - Soil Mixture

Fly ash (C) hydrates to form cementitious products similar to those produced during the hydration of Portland cement. This property makes it a very cost effective stabilization agent for a wide range of applications (Ferguson, 1993). Application of fly ash (C) treatment can be directed towards subgrade stabilization in pavement applications in order to enhance the bearing capacity and reduce the shrink and swell potential. Other application can be stabilization of high PI borrow soils for use in the embankment construction or in Backfill construction against retaining walls. In order to evaluate the stabilizing potential of fly ash, a low plasticity clayey sand (20% Kaolinite + 80% Sand) was prepared in order to simulate the composition of soils available in Texas; it will be referred to as "clayey sand (K)" Following tests were performed to evaluate the stabilization potential of fly ash (C)

Compaction: Harvard miniature samples were prepared for the determination of maximum dry density and optimum moisture contents. It was found that there is no significant difference in the maximum dry density with increasing proportion of fly ash (C) in the fly ash (C) - clayey sand (K) mix. The dry density increased to 21.0 kN/m^3 (131 pcf) with the addition of 30% fly ash (C) from 20.5 kN/m^3 (128 pcf) corresponding to clayey sand (K) with no fly ash(C). On the other hand, the optimum moisture content showed a decreasing trend with an optimum moisture content of 9% for 30% fly ash (C) in the mix.

Unconfined Compressive Strength: Samples were prepared in Harvard miniature molds and tested after 7, 28 and 56 days of curing. Three mixes were prepared by mixing 10%, 20% and 30% fly ash (C) by weight to the clayey sand (K). The results of this test are shown in Fig. 7. It can be observed that there is a gain in strength due to fly ash (C) stabilization. This is primarily due to the formation of various calcium silicate hydrates and calcium aluminate hydrates.

The strength increased 15 times when the clayey sand was stabilized with 30% fly ash (C) as compared to 10% flyash (C).

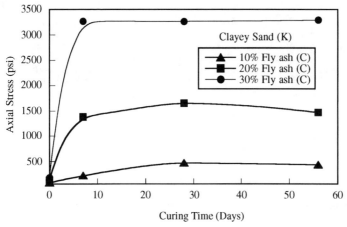

Figure 7. Variation of Unconfined Compressive Strength of Fly ash (C) Stabilized Clayey sand (K) with Curing Time

CBR: This test was conducted both on stabilized and unstabilized clayey sand (K). A proportion of 20% fly ash (C) was selected based on the economy of the stabilization process and also based on earlier studies (Nicholson, 1993). The CBR value for soaked clayey sand (K) was found to be 4 as compared to a CBR value of 1127 for stabilized clayey sand (K). This indicates a remarkable increase in the penetration resistance and it can be concluded that fly ash (C) stabilization is an effective means of enhancing the bearing capacity of clayey soils.

Consolidation: Consolidation characteristics determine the settlement potential of saturated soils. Samples having high compression index are susceptible to higher deformations for a given load. Compression index was determined for both stabilized and unstabilized clayey sand (K). For unstabilized soil, the compression index was 4% which decreased to 1% after stabilizing the soil with 20% Fly ash (C). This reduction is due to the hardening of the soil and increase in it's strength resulting from subsequent pozzolanic reactions.

Conclusions

Literature review indicates that three different types of embankment configurations have been used with recycled materials. Based on the experimental study, the following conclusions can be advanced:

1. According to USCS fly ash (C), fly ash (F), scrubber base (F), scrubber base (C), bottom ash and flourogypsum can be classified as either sand or silt. Tire chips can be characterized as a frictional material while all the other recycled materials exhibit frictional and cohesive properties (CD test).

2. Based on the CBR test results most of the materials tested were suitable for subgrade construction except for fly ash (F) and scrubber base (F) which had CBR values close to 1 under soaked condition. Hence fly ash (F) and scrubber base (F) cannot be used close to the watertable and proper drainage must be provided where ever these materials are used.

3. Environmental characterization indicates that all the selected materials are benign, considering the concentration levels of regulated metals in the leachate solutions. Sulfate concentrations in the leachate solutions of scrubber base (C), scrubber base (F) and flourogypsum were all higher than 1000 ppm which could be a problem for concrete structures in the neighborhood of the embankment due to sulfate attack.

4. Stabilization studies on a clayey sand (kaolinite based) showed that there is remarkable improvement in the strength and CBR value of fly ash(C) stabilized soil when compared to unstabilized soil.

Acknowledgment

This work was partly supported by the funding provided by the Texas Department of Transportation (TxDOT) to the Center for Innovative Grouting Materials and Technology (CIGMAT) at the University of Houston.

References

[1]   Ahmed, I., "Laboratory Study on Properties of Rubber Soils," Ph.D. Thesis, Purdue University, 1993, 450 p.

[2]   Ahmed, I., and Lovell, C. W., "Rubber Soils as Lightweight Geomaterials," Transportation Research Record No. 1422, 1993, pp. 61-70.

[3]   ASTM C 618-91, Standard Specification for Fly ash and Raw or Calcined Natural Pozzolan for Use as a Mineral Admixture in Portland Cement Concrete, ASTM, Philadelphia, PA.(www.astm.org)

[4]   Basheer, M., Vipulanandan, C. and O'Neill, M. W., "Recycled Materials in Embankments, Except Glass," Report No. CIGMAT/UH 96-1, University of Houston, Houston, Texas, 1996, 148 p.

[5]   Bloomquist, D., Diamond, G., Odeon, M., Ruth, B. and Tia, M.,"Engineering and Environmental Aspects of Recycled Materials for Highway Construction," FHWA Report, FHWA-RD-93-088, July 1993, 220 p.

[6]   Bosscher, P. J., Edil, T. B. and Eldin, N. N., "Construction and Performance of a Shredded Waste Tire Test Embankment," Transportation Research Record No. 1345, 1992, pp. 44-52.

[7]   Church, D. A., Raad, L. and Tumeo, M. "Experimental Study of Leaching of Fly ash," Transportation Research Record No. 1486, 1995, pp. 3-12.

[8]   Diamond, S., "On the Glass Present in Low-Calcium and in High-Calcium Fly Ash," Cement and Concrete Research, Vol. 13, No. 4, July 1983, pp. 459-464.

[9] DiGioia, A. M., Jr., and Nuzzo, W. L., "Fly ash as Structural Fill," Journal of Power Division, ASCE, Vol. 98, No. 1, 1972, pp. 77-92.

[10] Edil, T. B., Bosscher, P. J., and Eldin, N. N., "Development of Engineering Criteria for Shredded or Whole Tires in Highway Applications," Geotechnical Testing   Journal, Vol. 17, No. 4, 1992, pp. 453-464.

[11] EPA, "Markets for Scrap Tires," Report No. EPA/530-SW-90-074a, Washington, DC, October 1991.

[12] EPA, "Atomic Emission Spectrometric Method for Trace Element Analysis of Water and Wastes," EPA Method 200.7, 1982.

[13] EPRI, "Use of Ash in Highway Construction: Delaware Demonstration Project," EPRI Report GS-6540, November 1989, 70 p.

[14] Ferguson, G., "Use of Self Cementing Fly ashes as a Soil Stabilization Agent," Fly ash for Soil Improvement, ASCE Geotechnical Special Publication No. 36., 1993, pp. 1- 14.

[15] Gray, D. H. and Lin, Y. K., "Engineering Properties of Compacted Fly ash," Journal of Soil Mechanics and Foundation Division," ASCE Vol. 98, No. SM4, April, 1972, pp. 361-380.

[16] Hoppe E. J., "Field Study of a Shredded-Tire Embankment," Interim Report, FHWA/VA-94-IR1, June 1994, 46 p.

[17] Krizek, R. J., Chu, S. C. and Atmatzidis, D. K., "Geotechnical Properties and Landfilled Disposal of FGD Sludge," Geotechnical Practice for Waste Disposal, ASCE Geotechnical Special Publication No. 13, June 1987, pp. 625-639.

[18] Lovell, C. W., Chih Ke, T., Huang, W. H. and Lovell, J. E., "Bottom ash as a Highway Material, " Transportation Research Record No. 1310, 1991, pp. 106-116.

[19] Maryland DOT, "Fly ash Specification," Contract Q 627-501-270, 1993.

[20] McLaren R. J. and DiGioia A. M. "Typical Engineering Properties of Fly ash," Geotechnical Practice for Waste Disposal, ASCE Geotechnical Special Publication No. 13, June 1987, pp. 683-697.

[21] MPCA, "Waste Tires in Subgrade Road Beds," Minnesota Pollution Control Agency, St. Paul, MN, 1990, 34 p.

[22] NCHRP Synthesis 199, "Recycling and Use of Waste Materials and By-Products in Highway Construction," , Transportation Research Board, National Academy Press, Washington, D.C. 1994a.

[23] NCHRP Synthesis 198, "Uses of Recycled Rubber Tires in Highways," Transportation Research Board, National Academy Press, Washington, D.C., 1994b.

[24] Nicholson, P. G. and Kashyap, V., "Flyash Stabilization of Tropical Hawaiian Soils," Fly ash for Soil Improvement, ASCE Geotechnical Special Publication No. 36., 1993, pp. 15-29.

[25] Ormsby, W. C. and Fohs, D. G., "Use of Waste and By-Products in Highway Construction," Transportation Research Record No. 1288, 1990, pp. 47-58.

[26] Parker, D. G., Thornton, S. I. and Cheng, C. W., "Permeability pf Fly ash Stabilized Soils," Geotechnical Disposal of Solid Waste, ASCE Speciality Conference, 1977,   pp. 63-70.

[27] Roy D M., Luke, K., and Diamond, S., "Characterization of Fly ash and Its Reactions in Concrete," Proceedings, Materials Research Society, Pittsburgh, 1984, pp. 124-149.

[28] Read, J., Dodson, T, and Thomas, J., "Use of Shredded Tires for Lightweight Fill," Post Construction Report, Experimental Project, Oregon Department of   Transportation, Salem, Feb. 1991.

[29] Snow, P. G., Gehrmann, B. and Carrasquillo, R. L., "Production and Utilization of Scrubber Sludge," Proceedings: Eighth International Ash Utilization Symposium, EPRI CS-5362, Vol. 1, October 1987, pp. 6-1 to 6-13.

[30] Texas Recycles 2, Marketing Our Neglected Resources, Texas General Land Office, 1994, 106 p.

[31] Upton, R. J., and Machan, G., "Use of Shredded Tires for Lightweight Fill," Transportation Research Record No. 1422, 1993, pp. 36-45.

[32] Vipulanandan, C. and Basheer, M., "Fly ash and Tire Chips for Highway Embankments," Proceedings, Materials for New Millennium, ASCE, Washington, D.C. pp. 593-602, 1996.

UTILIZATION OF LAGOON-STORED LIME IN EMBANKMENT CONSTRUCTION

Khaldoun Fahoum,[1] Assoc. Member ASCE

Abstract

This paper summarizes a study conducted to evaluate the potential use of lagoon-stored lime as part of an embankment fill supporting a road extension crossing the existing lime lagoon. Approximately 150 meters of the proposed road extension will cross the existing lagoon which contains lime that was deposited after being utilized for water treatment in the nearby water treatment plant. The main focus of this study was to determine the maximum percentage of lime that can be mixed with a predetermined borrow material to build the proposed road embankment without adversely affecting the physical and mechanical properties of the borrow material. The borrow material consisted of high plastic clay classified as CH by the Unified Soil Classification System. Improvement of the soil properties such as reduction in its plasticity or increase in its strength, usually obtained upon mixing with active lime, was not expected due to the fact that the lime within the lagoon had already been utilized in the water treatment process. This was supported by test results on the lagoon lime and borrow materials which included Atterberg limits, compaction properties, unconfined compression, and unconsolidated-undrained (UU) tests. The tests were performed on soil-lime mix with lime percentages ranging from 0% to 40% of lime. The lime addition, acting as a lighter filler, affected the compaction properties of the soil by reducing the maximum dry density and slightly increasing the optimum moisture content. Compaction at 20% lime content was relatively difficult to control where a well-defined maximum dry density and optimum moisture content were not clear. Results of the Atterberg limits tests indicated a reduction in the plasticity indices of the soil by up to 22%. However, this reduction was not significant and the soil-lime mix was still classified as CH. Unconfined compression test data indicate a slight drop in undrained shear strength with the addition of lime. Data from the UU tests indicated a slight increase in the undrained shear strength with 10% lime and a smaller increase in strength with 20% lime. The observed fluctuation in strength may be attributed to several

---

[1] Senior Engineer, Geotechnology, Inc., 2258 Grissom Dr., St. Louis, Missouri 63146, Tel: (314) 997-7440, email: kf@geotechnology.com

factors including the effect of lime in lowering the density of the samples, sample preparation variations, and variations in the remolding water content. However, this fluctuation in strength is not very significant and is not expected to have a substantial effect on the engineering performance of soil-lime mixture. In conclusion, it appears that soil-lime mixture should perform satisfactorily as an engineered fill provided it is processed, mixed, and compacted properly. The maximum lime content should be kept below 15% in order to reduce problems associated with handling, proper mixing, and attaining the proper densities.

Introduction

The construction of a new road necessitates that the alignment of that road passes through an existing lagoon in which lime had been deposited from previous water treatment activities in a nearby water treatment plant. The road embankment is proposed to be built, within the lagoon area, using soil obtained from a predetermined borrow source. The lime within the embankment area is to be removed prior to construction. An evaluation of the potentials for utilization the maximum quantity of the previously deposited and removed lime in conjunction with embankment fill was conducted and is presented here. Chemical analysis of the lime was carried out by others. The information, analyses, and conclusions contained in this paper are based on information obtained from a geotechnical report prepared by Geotechnology, Inc. (Geotechnology, 1997) for this project.

Project Description

The total project includes construction of an approximately 600-meter extension of Elm Street to an intersection with the existing Mueller Road in St. Charles, Missouri. The northernmost 150 meters of the Elm Street extension will cross an existing lime lagoon, as shown in Figure 1. The lagoon is contained within an approximately 2-meter high earth berm. The lime inside the lagoon within the project area is approximately 1 to 1.5 meters deep. The proposed soil borrow area for the embankment construction, also shown in Figure 1, is located southwest of Mueller Road.

Field Investigation

The field investigation consisted of site reconnaissance and drilling two borings within the proposed borrow area in order to identify, classify, and sample the borrow material. The borings were drilled to a depth of 5 to 6 meters with a drill rig equipped with hollow-stem augers. Split-spoon as well as bulk samples were obtained from borings drilled using the drill rig. The stratigraphy, within the borrow area, consisted of one layer, below grass and top soil, of medium stiff to soft, light gray to gray, highly plastic clay extending to the depth of exploration.

Testing Program

The testing program was designed to investigate various properties of the natural borrow materials when mixed with different percentages of the lagoon lime. The percentages of the added lime were measured with respect to dry weight of soil. Soil-lime mix was cured for approximately 1 hour prior to testing.

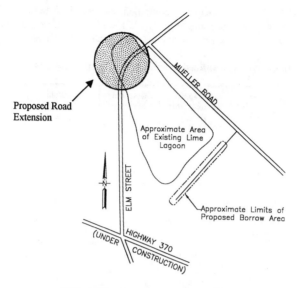

FIG. 1 Site Location and General Layout

Atterberg limit tests (ASTM D4318) were performed on mixtures of the natural soil with 0, 5, 10, 20 and 40 percent of lagoon lime, and with 5 percent of commercial active hydrated lime. Results of the Atterberg limit tests are listed in Table 1.

Table 1. Summary of Atterberg Limit Test Results

Additive	LL%	PL%	PI%	Classification
0% Lagoon Lime	94	33	61	CH Material
5% Lagoon Lime	81	26	55	CH Material
10% Lagoon Lime	84	28	56	CH Material
20% Lagoon Lime	80	28	52	CH Material
40% Lagoon Lime	79	28	51	CH Material
5% Commercial Active Lime	75	54	21	MH Material

Samples for the strength tests (unconfined and unconsolidated-undrained) were remolded and compacted at approximately 90% of their maximum modified Proctor dry density and at approximately 2% above their optimum moisture content to simulate practical field conditions. Samples were approximately 3.6 cm in diameter and 7.6 cm in height.   Saturation for the UU test samples was accomplished using back pressure of up to 275 kN/m². A summary of unconfined strength and UU test results are included in Tables 2 and 3, respectively.

Table 2. Summary of Unconfined Compression Strength Results

Additive % (Lagoon Lime)	Maximum Dry Density (kN/m³)	Dry Density As Compacted (kN/m³)	Optimum Moisture Content (%)	Moisture Content As Compacted (%)	Undrained Shear Strength (kN/m²)
0	16.2	14.9	20.8	22.5	57
10	16.1	14.9	23.3	24.6	46
20	15.5	14.4	22.0	24.1	50

Table 3. Summary of Unconsolidated-Undrained Test Results

Additive (Lagoon Lime) (%)	Maximum Dry Density (kN/m³)	Dry Density As Compacted (kN/m³)	Optimum Moisture Content (%)	Moisture Content As Compacted (%)	Moisture Content After Test (%)	Undrained Shear Strength (kN/m²)
0	16.2	14.1	20.8	25.4	40.9	7.2
10	16.1	14.5	23.3	27.4	39.0	12.4
20	15.5	14.1	22.0	24.2	42.6	9.1

Modified Proctor (ASTM D 1557), unconfined compression (ASTM D 2166), and unconsolidated-undrained (ASTM D 2850) tests were performed on mixtures of the natural soil with 0, 10, and 20 percent of lagoon lime. The modified Proctor curves are shown in Figure 2. Both the borrow soil and the lagoon lime had to be processed and dried prior to mixing and compacting. The natural borrow materials and the lagoon lime samples had moisture contents of approximately 43% and 110%, respectively.

Data Analysis
Compaction Tests. In general, the lagoon lime acted as lightweight filler for the soil mixture. The addition of the lime resulted in reductions in the maximum dry densities and small increases in the optimum moisture contents, as shown on Figure 2. This behavior is well documented in the literature and is attributed to the lower bulk density and the flocculating effects of the lime (Mitchell and Hooper, 1961; Neubauer and Thompson, 1972). The decrease in dry density and increase in optimum moisture content was not, however, as pronounced as with similar soils mixed with active lime (Fahoum and Aggour, 1995). This is mainly due to the fact that the lagoon lime had already been utilized previously and its flocculating characteristics have been reduced. The natural soil when mixed with 20% of lagoon lime was generally difficult to compact due to the increased percentage of the lighter

and finer lime particles within the mixture. The 20% lime sludge mixture had a resulting Proctor curve that does not have a well-defined maximum dry density and optimum moisture content.

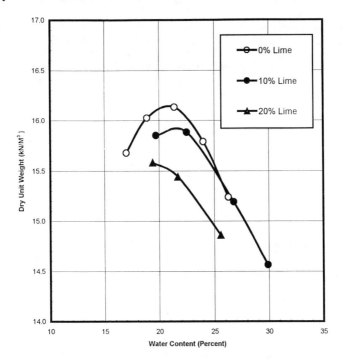

FIG. 2 Modified Proctor Density Curves

Atterberg Limit Tests. The results of the Atterberg limits tests, as shown in Table 1 indicate that the borrow soil has liquid limits in excess of 90% and plasticity indices in excess of 60%. Material with these Atterberg limits (classified as CH by the Unified Soil Classification System) are highly expansive and highly prone to excessive volume change upon wetting or drying. Based on correlations between plasticity index and swell potentials (Seed et al, 1962), the borrow soil is considered to have very high swell potential. Although the addition of the lagoon lime in percentages of 5, 10, 20, and 40, reduced the plasticity indices by 15 to 22 percent, it did not change the soil's expansive characteristics to a significant degree, as indicated by the same correlations, and the soil-lime mix was still considered to have very high swell potentials. Further, all of the soil-lime mixtures remained classified as highly plastic soil (CH). This leads to the conclusion that the lagoon lime did not react with the soil to reduce its expansive characteristics as would be expected with active lime. In general the addition of lime to cohesive soils would cause its liquid limit to decrease, the plastic limit to increase, and hence the plasticity index to decrease as a result of the cation exchange and the flocculation agglomeration reactions, as widely supported in the literature (Holtz, 1969; Jan and Walker, 1963; TRB, 1987; and Wang et al., 1963). The lime lagoon, being utilized for water treatment and stored under water for years, lost most of its activity and, therefore, did not behave as a stabilizer for the highly plastic borrow material. To further support this conclusion and to eliminate the possibility that the borrow soil was inactive, 5% of commercial active lime was mixed with soil. As shown in Table 1, a significant decrease in the plasticity index of approximately 68%, was achieved resulting in a mix that is considered to have low expansion potentials.

Strength Tests. Unconfined compression test data included in Table 2, indicate a slight drop in undrained shear strength with the addition of lagoon lime. Results obtained from the UU tests performed on saturated samples, as shown in Table 3, indicate some increase in the undrained shear strength with 10% lime and a smaller increase in strength with 20% lime. In general, this fluctuation in strength may be attributed to several factors including the effect of lime in lowering the density of the samples, accuracy in achieving the target densities, and sample preparation variation. Due to the fact that the lagoon lime has already been utilized, as indicated above, it appears that no pozzolanic reaction took place to significantly improve the strength of the treated borrow soil. The pozzolanic reaction is the reaction between the lime and the silica and alumina of soils to produce cementing materials. Significant increase in strength is normally achieved with the addition of active lime as a result of the pozzolanic reaction (Neubauer and Thompson, 1972; Thompson, 1966; and Thompson, 1966-a).

Conclusions

Based on the test results, it appears that soil-lime mixture would perform satisfactorily as an engineered fill provided it is processed, mixed, and compacted properly. This process will involve draining the lagoon within the project area and removing and drying the lime. The borrow soil also needs proper drying so that the soil-lime mix moisture content can be controlled. The soil and lime would then be mixed uniformly so that no more than 15% of lime, as measured with respect to the dry weight of soil, is used. The maximum recommended lime percentage is limited to 15% to reduce problems associated with handling, proper mixing, and attaining the proper densities. Compaction of the soil-lime mix should be such that it would produce minimum dry density of approximately 14.6 kN/m^3. The compaction moisture content is recommended to remain between 21 and 23 percent. Finally, the lagoon lime should not be expected to behave as soil stabilizer.

Acknowledgment

The author thanks the City of St. Charles, in particular Robert DeConcini, for granting permission to write this paper.

References

1.   Fahoum, K., and Aggour, M.S. (1995). "Range of Properties for Lime

Stabilized Clay Soils." Presented at the Transportation Research Board annual meeting in January in Washington D. C.

2.   Geotechnology, 1997, "Material Evaluation, Mueller Road and Elm Street Extension", St. Charles, Missouri, Report No. 3479.02.7111.

3.   Holtz, W.G., (1969), "Volume Change in Expansive Clay Soils and Control by Lime Treatment," Presented at the Second International Research and Engineering Conf. on Expansive Clay Soils, Texas.

4.   Jan, M., and Walker, R.D., (1963), "Effect of Lime, Moisture and Compaction on a Clay Soil," Highway Research Record 29, HRB, National Research Council, Washington, D.C., pp. 1-12.

5.   Mitchell, J.K., and Hooper, D.R., (1961), "Influence of Time Between Mixing and Compaction on Properties of a Lime-Stabilized Expansive Clay," Highway Research Board, Bulletin No. 304, National Research Council, Washington, D.C., pp. 14-31.

6.   Neubauer, C.H., and Thompson, M.R., (1972), "Stability Properties of Uncured Lime-Treated Fine Grained Soils," In Highway Research Record 381, HRB, National Research Council, Washington D.C., pp. 20-26.

7.   Seed, H.B., Woodward, R.J., and Lundgren, R., (1962), "Prediction of Swelling Potential for Compacted Clay,", Proc. ASCE, JSMFD, Vol. 88, No. SM3, pp. 53-87.

8.   Thompson, M.R., (1966), "Shear Strength and Elastic Properties of Lime-Soil Mixture," In Highway Research Record 139, HRB, National Research Council, Washington D.C., pp. 1-14.

9.   Thompson, M.R., (1966-a), "Lime Reactivity of Illinois Soils," Proc. ASCE, JSMFD, Vol. 92, No. SM5, pp. 67-92.

10.   Transportation Research Board, (1987), "Lime Stabilization," State-of-the-Art Report 5, TRB, National Research Council, Washington, D.C., pp. 3-5.

11.   Wang, J.W.H., Mateos, M., and Davidson, D.T., (1963), " Compaction Effects of Hydraulic Calcitic and Dolomitic Limes and Cement in Soil Stabilization," Highway Research Record 29, HRB, National Research Council, Washington, D.C., pp. 42-54.

Behavior of Construction and Demolition Debris in Base and Subbase Applications

W. J. Papp Jr.[1], M. ASCE, M. H. Maher[2], M. ASCE, T.A. Bennert[3], M. ASCE and N. Gucunski[4], M. ASCE

Abstract

The design performance of a pavement system is evaluated based on the predicted number of load repetitions it will be subjected to throughout its service life. Mechanistic pavement design predicts pavement failure due to cracking and/or rutting. Cracking of the pavement is induced by tensile strains at the bottom of the asphalt layer, while rutting may be a result of accumulated permanent deformation throughout the pavement system. The pavement system includes a bound bituminous layer and an unbound base, subbase, and subgrade layer. The amount of base and subbase material in their respective quarries and borrow pits are of finite supply. As the production of demolition and construction debris increases and the availability of landfill space decreases, alternative methods of disposal are needed. A viable solution for the disposal of these materials is to use them in base and subbase applications.

A laboratory investigation was conducted on two types of construction and demolition debris, recycled asphalt pavement (RAP) and recycled concrete (RCA).

[1]Graduate Research Assistant, Department of Civil and Environmental Engineering, Rutgers, The State University of New Jersey, Piscataway, NJ 08854-8014

[2]Associate Professor and Chair, Department of Civil and Environmental Engineering, Rutgers, The State University of New Jersey, Piscataway, NJ 08854-8014

[3]Graduate Research Assistant, Department of Civil and Environmental Engineering, Rutgers, The State University of New Jersey, Piscataway, NJ 08854-8014

[4]Associate Professor, Department of Civil and Environmental Engineering, Rutgers, The State University of New Jersey, Piscataway, NJ 08854-8014

The recycled materials were compared to dense graded aggregate (DGA), which is currently used as a base material in New Jersey. Evaluation of permanent deformation and resilient modulus were conducted in the lab by cyclic load triaxial tests. The test results indicated RAP and RCA yielded higher resilient modulus values than DGA. Permanent deformation results indicated RCA experienced the least amount of permanent deformation. In contrast, RAP developed the largest amount of permanent deformation with respect to DGA. Existing models currently used for virgin materials were utilized to predict resilient modulus and permanent deformation in the recycled materials. Laboratory test results concluded these models could be used for evaluating resilient modulus and permanent deformation in unbound recycled materials.

Introduction

Recycling of construction and demolition debris has been increasing at an elevated rate, with little long-term research conducted on the durability of these materials. Potential savings in construction cost and time has made the use of RAP and RCA an attractive alternative to the highway engineer (Maher et al., 1997). To utilize large volumes of construction and demolition debris, the minimum standards set by AASHTO, as well as the state specifications, must be met. Since these recycled materials may be generated on the job site, the quality control of these recycled materials may change during the course of the project. Therefore, highway engineers choose to blend the recycled materials with quarried DGA, which consistently meets the AASHTO and state specifications. This paper reports the material properties of 100% RAP, RCA, DGA and blends of RAP and RCA mixed with DGA.

In general, static testing procedures are not adequate for determining the behavior of soil and aggregate materials subjected to the impulse type repeated loading representative of moving wheel loads (The Aggregate Handbook, 1993). In recent years, pavement engineers have recognized cyclic load triaxial tests as a fundamental mechanism for evaluating aggregate base and subbase soils. The cyclic load triaxial test provides a means of representing traffic type loading, as well as measuring the resilient and permanent strains that occur due to cyclic loading (Barksdale, 1972; Khedr, 1985).

Background

Resilient Modulus

Resilient Modulus (M_r) is an index that describes the nonlinear stress-strain behavior of soils under repeated loads. Resilient modulus is defined as the ratio of cyclic deviator stress to the magnitude of recoverable strain for a given cycle (Figure 1). Resilient modulus is expressed as the following:

$$M_r = \sigma_{cd} / \varepsilon_r \qquad (1)$$

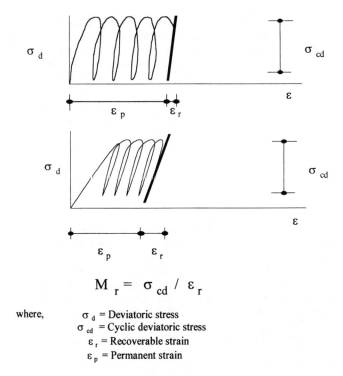

$$M_r = \sigma_{cd} / \varepsilon_r$$

where, σ_d = Deviatoric stress
            σ_{cd} = Cyclic deviatoric stress
            ε_r = Recoverable strain
            ε_p = Permanent strain

Figure 1 – Schematic of Theoretical Resilient Modulus and Permanent Deformation

where,

        M_r = resilient modulus
        σ_{cd} = cyclic deviatoric stress
        ε_r = recoverable strain

In general, a number of factors affect resilient modulus, some of which are: moisture content, density, stress history, aggregate type, gradation, temperature, % fines, and degree of saturation. Some models previously developed include such material parameters as cohesion, friction angle, bulk stress, major principal stress, and moisture content (Bonaquist and Witczak, 1996; Tian et al., 1998). However, in this study, the AASHTO bulk stress model was used to predict the resilient modulus for RAP, RCA, DGA, and the blended recycled material. This model can be expressed as the following:

$$M_r = k_1 \Theta^{k2} \qquad\qquad (2)$$

where,

M_r = resilient modulus
k_1 & k_2 = material parameters based on laboratory results
Θ = bulk stress ($\sigma_d + 3\sigma_3$)

It also must be noted that, although considerable emphasis has lately been placed on resilient moduli, from a practical viewpoint, the permanent deformation characteristics are often more important since it is the permanent deformation that results in the failure of the pavement system (Thompson and Smith, 1990; The Aggregate Handbook, 1993).

Permanent Deformation

In mechanistic-empirical methods of design, the material properties of the pavement base, subbase, and subgrade are essential in analyzing the response of the pavement system to vehicular loading. By knowing these properties, the stresses and strains developed in the material can be determined to predict the failure of the pavement system. For flexible pavements, the prediction of failure is based on determining the fatigue cracking, rutting, and low temperature cracking of the pavement system (Huang, 1993). Fatigue cracking of a pavement system is the result of horizontal tensile strains (ε_t) at the base of the asphalt layer, while rutting is the result of vertical compressive strains (ε_{vc}) that originate at the asphalt layer and migrate to the subgrade (Figure 2). Due to the vertical compressive strains, permanent deformation occurs throughout the pavement system.

To predict permanent deformation in a pavement system, cyclic load triaxial tests of granular materials have been the most widely used and recognized. Barksdale (1972) and Lentz and Baladi (1981) have shown that the permanent strain at a given cycle is proportional to the permanent strain at the first cycle plus the logarithm at that given cycle. This relationship can be expressed as the Log N model (Equation 3).

$$\varepsilon_p = a + b \log N \tag{3}$$

where,

ε_p = accumulated plastic strain
N = number of loading cycles
a = permanent deformation of the first cycle
b = slope of the least squares regression analysis

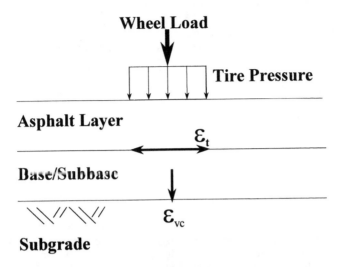

Figure 2 – Schematic of Mechanistic-Empirical Failure Criteria

On the other hand, Diylajee and Raymond (1982) and Vuong and Armstrong (1991) expressed permanent deformation as the following Exponential model:

$$\varepsilon_p = aN^b \qquad (4)$$

where,

ε_p = accumulated plastic strain
N = number of loading cycles
a = permanent deformation of the first cycle
b = slope of the least squares regression analysis

Both equations (3) and (4) will be implemented and compared to determine which model is more suitable for predicting the permanent strain in the construction and demolition debris.

Laboratory Testing Program

A laboratory testing program was established to determine the resilient modulus and permanent deformation of RAP, RCA, DGA and DGA blended recycled material. Grain size analysis, as well as standard compaction tests were also conducted to determine the basic physical properties of the material.

The testing system used for the resilient modulus and permanent deformation testing was an MTS closed loop servo-hydraulic testing frame (Figure 3). A load cell with a capacity of 15 kN was used for accurate load measurements. The loading frame is equipped with a hydraulically controlled

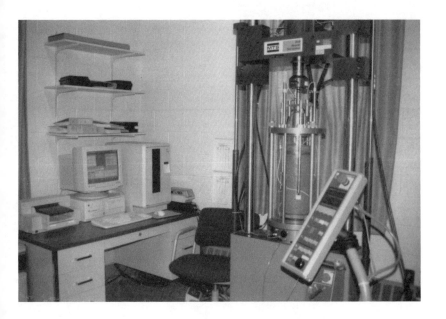

Figure 3 - Testing Frame, Triaxial Cell, and Data Acquisition System

movable crosshead to allow easy placement of the triaxial chamber. The top plate of the chamber was modified to hold a pressure transducer and four LVDTs. The two of the LVDTs measured the axial deformations of the specimen, as required by AASHTO TP46-94.

The system was programmed to record readings from the two LVDTs and the pressure transducer every time a peak and valley occurred from the load cell. The data was acquired through the MTS Teststar II System and saved in a spreadsheet format.

Test Results

Grain Size Analysis

Grain size analysis was utilized to determine the grain size characteristics of the construction and demolition debris.   Figure 4 shows the grain size distribution for the three materials used in the research program.

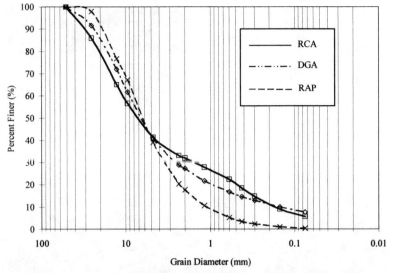

Figure 4 - Grain Size Distribution for the Construction/Demolition Debris

Moisture-Density Relationship

Standard compaction tests were conducted in accordance to ASTM D698 to determine the maximum dry unit weight and optimum moisture content for all samples used in the resilient modulus and permanent deformation tests. A summary of the compaction tests is shown in Table 1.

Table 1 – Summary of Maximum Dry Density and Optimum Moisture Content from Standard Compaction Tests

Material	Maximum Dry Density (kg/m³)	Optimum Moisture Content (%)
100% DGA	2098	7.0
75% DGA, 25% RAP	2056	7.0
50% DGA, 50% RAP	2028	6.0
25% DGA, 75% RAP	1956	5.5
100 % RAP	1872	5.0
75% DGA, 25% RCA	2064	5.0
50% DGA, 50% RCA	2016	6.0
25% DGA, 75% RCA	2000	7.0
100% RCA	1984	7.5

Resilient Modulus

Resilient modulus testing was conducted in accordance with AASHTO designation TP46-94 on RAP, RCA, DGA, and DGA blends. Each sample was compacted to the maximum dry density by means of a vibratory compaction hammer.

Resilient modulus, as a function of bulk stress for the 100 % and blended samples, is shown in Figures 5 and 6. The 100 % RAP sample produced the highest resilient modulus, while the 100 % DGA sample produced the lowest. Each blended sample, both the RCA and RAP, were strongly influenced by the percent of DGA added. As shown in both figures, the resilient modulus decreased as the percentage of DGA increased. Table 2 compares the resilient modulus at bulk stresses of 138 kPa and 344 kPa for all blends, as well as the resilient modulus of the blended samples to 100 % DGA.

To verify equation (2), regression analysis was performed to determine material parameters, k_1 and k_2. Predicted resilient modulus was determined at bulk stresses of 138 kPa and 344 kPa with the corresponding material parameters, k_1 and k_2. The percent differences between the predicted and actual values were calculated and presented in Tables 3 and 4. As shown in the table, the resilient modulus values predicted by equation (2), produced minimal percent differences to the actual values for the RAP, RCA, DGA, and DGA blended material.

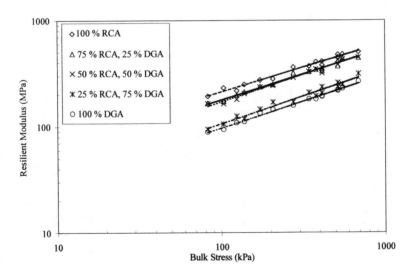

Figure 5 - Resilient Modulus for RCA Blended Samples

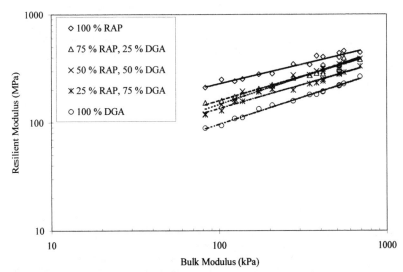

Figure 6 - Resilient Modulus for RAP Blended Samples

Table 2 – Comparison of Resilient Modulus Results at Bulk Stresses of 138 kPa
and 344 kPa

Material	M_r @ bulk stress of 138 kPa (MPa)	M_r @ bulk stress of 344 kPa (MPa)	% Increase in M_r @ bulk stress of 138 kPa compared to DGA	% Increase in M_r @ bulk stress of 344 kPa compared to DGA
100% DGA	111	157	0	0
75% DGA, 25% RAP	157	288	41.4	83.4
50% DGA, 50% RAP	194	277	74.7	76.4
25% DGA, 75% RAP	183	274	64.8	74.5
100 % RAP	253	342	127.9	117.8
75% DGA, 25% RCA	119	209	7.2	33.1
50% DGA, 50% RCA	205	327	84.7	108.3
25% DGA, 75% RCA	210	328	89.2	108.9
100% RCA	249	370	124.3	135.7

Table 3 – Comparison of Actual and Predicted Resilient Modulus at Bulk Stresses of 138 kPa.

Material	M_r @ 138 kPa Act.	M_r @ 138 kPa Pred.	% Diff. from Act.	k_1	k_2
100% DGA	111	113	1.8	9.553	0.502
75% DGA, 25% RAP	157	156	0.64	17.35	0.456
50% DGA, 50% RAP	194	188	3.09	14.58	0.519
25% DGA, 75% RAP	183	185	1.09	19.49	0.457
100 % RAP	253	259	2.37	43.10	0.364
75% DGA, 25% RCA	119	125	5.04	9.746	0.518
50% DGA, 50% RCA	205	201	1.95	16.12	0.512
25% DGA, 75% RCA	210	209	0.48	19.26	0.484
100% RCA	249	246	1.2	25.35	0.462

Table 4 – Comparison of Actual and Predicted Resilient Modulus at Bulk Stress of 344 kPa.

Material	M_r @ 344 kPa Act.	M_r @ 344 kPa Pred.	% Diff. from Act.	k_1	k_2
100% DGA	157	179	14.01	9.553	0.502
75% DGA, 25% RAP	288	234	18.75	17.35	0.456
50% DGA, 50% RAP	277	302	9.03	14.58	0.519
25% DGA, 75% RAP	274	281	2.55	19.49	0.457
100 % RAP	342	361	5.56	43.10	0.364
75% DGA, 25% RCA	209	201	3.83	9.746	0.518
50% DGA, 50% RCA	327	321	1.83	16.12	0.512
25% DGA, 75% RCA	328	325	0.91	19.26	0.484
100% RCA	370	376	1.62	25.35	0.462

Permanent Deformation

Testing was conducted on RAP, RCA, DGA, and DGA blends to determine the permanent deformations and strains in the respective material. Each sample was compacted to the maximum dry density by means of a vibratory compaction hammer. A constant confining stress of 103.42 kPa was applied to each sample during the cyclic test. The samples were loaded under a cyclic deviatoric stress of 310.26 kPa until 100,000 repetitions were achieved. Figures 7 and 8 show the accumulation of permanent strain for both the RCA blend, and the

RAP blend, respectively.   As shown from these figures, the accumulation of permanent strain is the lowest for the 100 % RCA and the highest for the 100 % RAP. The figures also show that the accumulation of permanent strain for the blend material is a function of the percent of DGA added to the blends. Table 5 provides a summary of the models which best predicted the permanent strain for all the samples tested.   The table also provides the percent difference of the predicted permanent deformation to the actual permanent deformation.

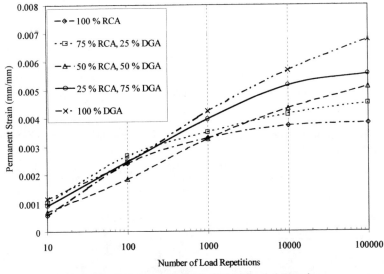

Figure 7 – Permanent Strain for RCA Blended Samples

Based on Table 5, Figure 9 was produced to illustrate the accumulation of permanent strain for the actual and predicted values of the 100 % samples. As presented in Figure 9, the Exponential model for the 100 % RAP adequately predicts the permanent strain of RAP until 10,000 cycles. At this point, the model over-predicts the actual permanent strain accumulating in the sample. Meanwhile, the lower strained samples, 100 % DGA and 100 % RCA, were notably similar to the Log N model.

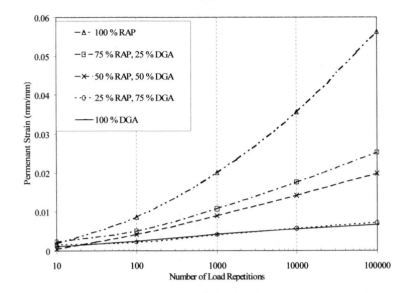

Figure 8 – Permanent Strain for RAP Blended Samples

Table 5 – Model Type Determined to Best Predict the Permanent Strain in
Construction/Demolition Debris and the Percent Difference from the Actual

Material	Model Type Used	Percent Difference from the Actual (%)
100% DGA	Log N Model	3.5
75% DGA, 25% RAP	Log N Model	11.23
50% DGA, 50% RAP	Log N Model	16.81
25% DGA, 75% RAP	Exponential Model	13.81
100 % RAP	Exponential Model	30.69
75% DGA, 25% RCA	Log N Model	15.77
50% DGA, 50% RCA	Log N Model	6.31
25% DGA, 75% RCA	Log N Model	16.19
100% RCA	Log N Model	31.81

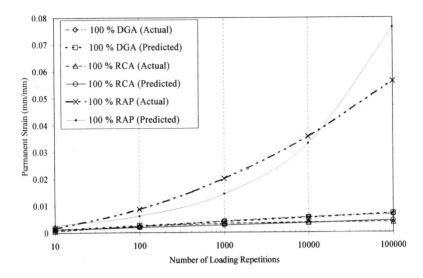

Figure 9 – Predicted Permanent Strain of the 100 % Samples Versus the Actual
Permanent Strain

Conclusions

Resilient modulus and permanent deformation testing was used to determine the cyclic response of construction and demolition debris under traffic type loading for pavement base and subbase conditions. From the testing conducted, the following conclusions can be made:

1. The 100 % RAP sample obtained the greatest value for resilient modulus. However, the 100 % RAP accumulated the greatest amount of permanent strain. The degradation of the asphalt binder may have attributed to the elevated values of permanent strain.
2. The material which performed the best when combining the characteristics of the resilient modulus properties as well as permanent strain, was the 100 % RCA. This material experienced the least amount of permanent strain while providing the second largest value for resilient modulus.
3. The Bulk Stress model, equation (2), typically used for predicting the resilient modulus of virgin soils, proved to provide an accurate representation of the resilient properties of construction and demolition debris under traffic type loading conditions.
4. Both the Log N model and the Exponential model were successfully fitted to the RAP, RCA, DGA, and DGA blended material. However, each respective

model was not successfully fitted to each sample. The model that best predicted the permanent strain for each sample is shown in Table 5.
5. As the percent of DGA increased in the RAP and RCA blended samples, the resilient modulus values decreased. However, the addition of DGA in RAP blended samples provided a decrease in the accumulation of permanent strain. Unfortunately, the accumulation of permanent strain increased when the DGA was added to the RCA blended samples.
6. The use of construction and demolition debris in the base and subbase layers of pavement systems can be a viable and cost effective material for pavement designs.

Acknowledgement

The research presented in this paper was funded by the New Jersey Department of Transportation. The authors wish to thank Mr. Frank Palise, Nicholas Vittilo, and Victor Mottola for their input regarding this paper.

References

Barksdale, R.D., (1972), "Repeated Load Test Evaluation of Base Course Materials," Georgia Highway department Research Project 7002. Georgia Institute of Technology, Atlanta.

Bonaquist, R. and Witczak, M.W., (1996), "Plasticity Modeling Applied to the Permanent Deformation Response of Granular Materials in Flexible Pavement Systems," Transportation Research Record No. 1540, Transportation Research Board, Washington, D.C. pp. 7-14.

Diylajee, V.A. and Raymond, G.P., (1982), "Repetitive Load Deformation of Cohesionless Soil," Journal of the Geotechnical Engineering Division, ASCE, Vol. 108, No. 10, pp. 49-56.

Huang, Yang, H., Pavement Analysis and Design, Prentice-Hall, Englewood Cliffs, New Jersey, 1993.

Khedr, S., (1985), "Deformation Characteristics of Granular Base Course in Flexible Pavements," Pavement System Analysis, Transportation Research Record No. 1043, Transportation Research Board, Washington, D.C. pp. 131-138.

Lentz, R.W. and Baladi, G.Y., (1981), "Constitutive Equation for Permanent Strain of Sand Subjected to Cyclic Loading," Transportation Research Record No. 810, Transportation Research Board, Washington, D.C. pp. 50-54.

Maher, M.H., Gucunski, N., Papp Jr., W.J., (1997), "Recycled Asphalt Pavement as a Base and Sub-Base Material," Testing Soil Mixed with Waste or Recycled Materials, ASTM STP1275, American Society for Testing and Materials, Philadelphia pp. 42-53.

The Aggregate Handbook, National Stone Association, Washington D.C., 1993.

Thompson, M.R. and Smith K.L., (1990), "Repeated Triaxial Characterization of Granular Bases," Dynamic Testing of Aggregates and Soils and Lateral Stress Measurements," Transportation Research Record No. 1278, Transportation Research Board, Washington, D.C. pp. 7-17.

Tian, P., Zaman, M.M., Laguros, J.G., (1998), "Gradation and Moisture Effects on Resilient Moduli of Aggregate Base," 77th Annual Transportation Research Board Meeting, January 11-15, Washington, D.C.

Vuong, B. and Armstrong, P., (1991), "Repeated Load Triaxial Testing on the Subgrades from Mulgrave ALF Site," WD RI91/023. Australian Road Research Board.

Role of Constituents on the Behavior of Flowable Fly Ash Fill

C. Vipulanandan[1] M.ASCE, Y. Weng[2] and C. Zhang[2]

Abstract

Flowable fills are increasingly used as backfill materials for pipes and retaining structures because of ease of placement and performance of the solidified material. But the role of its constituents on the working and mechanical properties of the flowable fill are not well understood. In this study, flowability, setting time and mechanical properties of various constituents of the flowable fills were investigated. Cement required almost double the quantity of water compared to fly ash (Class-F) to achieve similar flowability (Flow Table method). The Behavior of two fly ash (Class-F) rich flowable fill mixtures with different cement contents (1.5 and 3%) was investigated. The 28-day compressive strength for the two mixtures were 1.6 and 5.7 MPa. Several methods were used to quantify the flowability and setting time of the flowable fills. Effects of cement on the compressive and tensile strengths of the solidified flowable fill were investigated. The relationship between tensile strength and compressive strength of the binders and the flowable fills were linear and the strength ratios varied between 0.11 and 0.16.

Introduction

Flowable fill is a self-leveling, high slump mixture of Portland cement, fly ash, sand and water with selected additives. It is also classified as a controlled low strength material (CLSM) by the ACI Committee 229 if it has a 28-day compressive strength not exceeding 8.3 MPa (1200 psi) (ACI, 1997). This material has several other generic names including lean-mix backfill, flowable mortar and controlled-density fill [Smith, 1991]. Flowable fill is not a concrete nor a soil-cement but it has properties similar to both. Flowable fills are increasingly used as backfill

---

[1]Professor and Director and [2]Graduate Student, Center for Innovative Grouting Materials and Technology (CIGMAT), University of Houston, Houston, Texas 77204-4791. Phone: (713)743-4278; email: cvipulanandan@uh.edu

materials for pipes and retaining structures because of ease of placement and performance of the solidified material. The consistency of flowable fill used in geotechnical applications is similar to a lean grout or slurry, yet several hours after placement the material hardens enough to support traffic loads without settlement. Typical 28-day compressive strengths range from   0.34 to 1.4 MPa (50 to 200 psi) which is more than adequate when compared to most compacted soil or granular fills [Ramme and Naik, 1998]. Unit weights of the flowable fills vary from 18 to 23 kN/m^3 [Ramme and Naik, 1998]. The flowable fill contains many of the same constituents found in concrete but in different proportions and hence it can be batched and mixed using the same equipment used in producing concrete and delivered to the job site by ready mix truck.

Applications

Flowable fill can replace compacted soil as structural fill or backfill in many applications [ACI, 1997; Landwermeyer et al. 1997]. Due to its flowability and self-leveling properties it requires no compaction and hence ideal for use in tight or restricted-access areas where placing and compacting soil or granular fill is difficult or even impossible. It is used to fill voids under existing pavements, buildings or other structures, backfilling narrow trenches and fill abandoned underground structures such as culverts, tunnels, storage tanks, wells and sewers.

Utility companies often specify flowable fill instead of soil for backfilling around pipes or conduits[Smith, 1991; Ramme and Naik, 1998]. The material flows under and around pipes providing uniform support without leaving voids. Self-leveling eliminates the chance of workers/equipment accidentally damaging pipes since compaction can be eliminated. Since easy access to utility lines are essential for maintenance and repair, flowable fill strength can be specified at or below 0. 7 MPa (100 psi). At this strength, the material is excavatable with backhoe or other digging equipment. Flowable fills are also becoming the choice for plastic pipes used in wastewater and stormwater applications.

Flowable fills are also used for pavement construction and maintenance [Adaska, 1994; ACI, 1997]. Used under roadways, it serves as a strong, stable subbase. As a fill material for pavement section replacement, the material solidifies enough to support the patch as soon as three hours after placement, depending on the mix used, weather conditions and the depth of trench. Increasing number of State Departments of Transportation (DOTs) are using/specifying flowable fill in their construction, repair and maintenance operations.

For most applications flowable fill can be placed continuously. Sometimes, though, it is necessary to place the material in lifts. When backfilling retaining walls, placing flowable fill in lifts prevents lateral pressure from exceeding the loading capacity of the wall. Allow each lift to harden before placing the next lift. For pipe bedding, placing in lifts prevent floating the pipe. Due to its unit weight the flowable fill displace the water in the trench. Because of its high water content flowable fill will bleed. Generally the bleed water is usually not a problem and can be allowed to run off or evaporate. When placing flowable fill in open trenches in cold weather, heat it using the same methods for heating ready mixed concrete to prevent the material from freezing before it hardens.

Recent innovations include the use of recycled materials such as glass and foundry sand, increased air content, color for rapid identification of buried utilities, blended coal ash and locally available aggregates [Naik et al. 1997; Ramme and Naik, 1998].

Costing about two-thirds to three-quarters the price of standard ready mixed concrete, flowable fill is expensive per cubic yard than most soil or granular fills. Still, the advantages of using flowable fill more than compensate for its higher cost. The cost depends on material used, mixing and transport, and placing method.

Testing

Four standards (3 testing and 1 practice) have been developed by ASTM for flowable fills (CLSM)[www.astm.org]. The standard tests are for flow consistency (D 6103), unit weight and air content (D 6023), and suitability for load application (D 6024). These test methods are under the jurisdiction of ASTM Committee D-18 on Soil and Rock and is the direct responsibility of Subcommittee D18.15 on Stabilization with Admixtures. The standard practice is for sampling freshly mixed CLSM (D 5971). Using standard slump cone test to verify flowability is not very accurate. ASTM C 403 penetration resistance test can assess the setting time and early strength development of flowable fill. Many other procedures for flowable fill follow the same ASTM standards used to test concrete.

Although many city public work departments, utility companies, and state departments of transportation have been backfilling with flowable fill since 1970s, no universal standards were developed until recently.

Mix Design and Performance

A typical flowable fill mix contains cement, fly ash, fine aggregates and water. In order to meet specific performance requirements these components can be mixed in varying proportions by taking advantage of the locally available materials. Screening locally available materials for use in the flowable fill is considered a challenge and no approach is available at this time. The objective of mix design will be to develop an economical mix with required properties. Hence the basic role of each component in the flowable fill must be understood [Janardhanam et al. 1992].

(i) Cement

As in cement concrete, the Portland cement (ASTM C 150) in flowable fill combines with the water to bind the aggregates and fly ash. Although flowable fill contains much smaller amount of cement than concrete, sufficient hydration occurs to produce a solid mass. Bleeding of cement within the mix must be minimized.

(ii) Fly ash

Fly ash is produced during the combustion of coal (ASTM C 618). Approximately 75 million tons of coal ash is produced in the U.S. annually (NCHRP, 1994a). The chemical compositions and physical appearance vary with the coal properties, burning processes, and collection procedures. Its principal constituents are silica, alumina, iron oxide, lime and carbon. According to ASTM C 618, fly ash is classified into Class-C and Class-F based on the content of calcium oxide (CaO) and other metal oxides. If the fly ash has a CaO content less than 10%, it is technically considered Class-F; otherwise, Class-C. Class-F fly ash is produced by burning anthracite or bituminous coal while Class-C fly ash is produced by burning lignite or sub-bituminous coal. The principal active component is siliceous or aluminosilicate glass for Class-F fly ash, and calcium aluminosilicate glass for Class-C fly ash (Diamond 1983, Roy, Luke; and Diamond, 1984).

The primary role of fly ash in flowable fill is to improve flowability. It also increases strength and reduces bleeding, shrinkage and permeability.

(iii) Sand

Sand increases the unit weight of the flowable fill but reduces its flowability. Generally flowable fill is more economical to produce with locally available sand. Various sands including pea gravel, sandy soil with not more than 10% passing sieve no. 200, quarry waste products, and ASTM C 33 specification aggregates have been used.

(iv) Water

Large quantity of water is used in flowable fill allowing the material to flow readily, self-consolidate and self-level. Water contents vary depending on flowability and strength requirements. Increasing the water-to-binder ratio increases flowability but reduces its compressive strength.

In this study, two fly ash modified flowable fill mixtures were investigated to determine the role of cement and other constituent on the flowability and mechanical properties. Several methods (including ASTM) to quantify the flowbility and setting time of the flowable fill were investigated. Effect of cement and curing time on the compressive and tensile strengths of the solidified flowable fills were investigated.

Objectives

The specific objectives of this study are as follows: (1) to characterize the behavior of constituents of flowable fills; and (2) to determine role of cement on the behavior of flowable fly ash fills.

Experimental Program

Materials

(i) Binders: The pH of the cement (Type I), fly ash-C and fly ash-F used in this study were 12.0, 11.23 and 10.74 respectively. The tests were performed according to ASTM D 5239.

(ii) Flowable fill : Table 1 summarizes the constituents and proportions used in the two flowable fill mixtures investigated in this study. The cement content in Mix-1 and Mix-2 were 3 and 1.5 percent respectively. In Mix-1 and Mix-2, fly ash (F)-to-cement ratios were 4 (11.9%) and 9 (13.5%) respectively with a water-to-binder ratio of 0.7 (10.4%). The components were first mixed by hand and then transferred to a floor model mixer to be blended with water. Initial mixing of the constituents was done in small batches in order to ensure good mixing. The sequence of mixing was as follows: cement and fly ash (F) were first hand mixed with sand and then water was added and mixed again in the Floor-Mixer. The working properties of flowable fill such as flowability and setting time were determined using standard ASTM tests.

Table 1. Material Composition, Flowability and Bleeding of the Two Test Mixtures.

Mix	Cement	Fly ash (F)	Sand: Binder	Water: Binder	Flowability (%)		Bleeding (%)
					Slump (in)	Flow Table	
1	1.0	4.0	5.0	0.7	100 (8)	80	1.6
2	0.5	4.5	5.0	0.7	100(8)	80	2.7

Methods

Working Properties

(a) Flowability

Several instruments were used for determining the flowability of cement, fly ashes and flowable fills. The methods are the Flow table, Flow cylinder and Slump test.

(i) Flow Table (ASTM C 230): In this test the table is raised and dropped 10 times in 6 seconds by rotating a hand wheel. Flowability is calculated as $100(d-4)/4$. Average weight of the sample tested each time was about 500 grams.

(ii) Flow Cylinder (ASTM D 6103-97/PS 28-95): The procedure consists of placing a 3 in. diameter x 6 in. long open ended cylinder vertically on a level surface and filling the cylinder to the top with grout. The cylinder is then lifted vertically, to allow the material to flow out onto the level surface. Good flowability is achieved where there is no noticeable segregation and the material spread is at least 8 in. diameter. Flowability is defined as $100 (d-3)/5$, where d is the diameter of material spread. Average weight of the sample tested each time was about 1300 grams.

(iii) Slump Test (ASTM C 143): Consistency of concrete is measured using this method. The consistency is described interms of centimeters of slump. When the slump is 200 mm (8 in.) the flowability is considered to be 100%.

(b) Setting time

Different instruments were used for determine the setting time of the binders (cement and fly ash) and flowable fill. The methods are the Vicat needle and penetration test.

(i) Needle Test (ASTM C 191): The Vicat's needle was used to determine the initial and final setting times of cement and fly ashes. The penetration of the 1.0 mm diameter needle was monitored with time. By definition, the initial time of set is the time corresponding to a needle penetration of 25 mm and the final time of set is the time corresponding to a needle penetration of less than 1 mm.

(ii) ***Penetration Test (ASTM C 403-95):*** The penetration resistance is used to determine the initial and final setting times of cement grouts. Initial and final setting times are defined as the elapsed time to reach a resistance of 500 and 4000 psi respectively.

(c) Mechanical Properties

(i) Unconfined Compressive Strength (ASTM C 109): The cylindrical specimens for uniaxial compression tests were 38 mm (1.5 in.) in diameter with height varied from 65 to 90 mm. Compression tests were performed using a screw-type machine with a capacity of 10 kips. The displacement rate was kept constant at 0.031 mm/min. The specimens were trimmed and capped to ensure parallel surfaces. At least three specimens were tested under each condition.

(ii) Indirect (Splitting) Tension Test (ASTM C 496): Tests were done on specimens with diameters of 38 mm (1.5 in.) and length varying from 65 to 90 mm. The screw-type machine of 10 kips capacity was used for determining the splitting tensile strength of grouts. At least three specimens were tested under each condition.

Results and Discussions

(a) Flowability

Dry Materials: Tests were performed on dry materials to evaluate their flowability and to determine whether materials could be selected based on these results. Both Flow Table (ASTM C 230) and Flow Cylinder (ASTM D 6103) tests were performed. Flowability of cement, fly ashes (Class C and Class F) and sand are shown in Fig. 1. Based on the Flow Table test, cement and both fly ashes show flowability of slightly over 100%. Based on the Flow Cylinder test, Fly ash (Class- F) had the best flowbility of the three binders and cement had the lowest flowability. Based on these limited test results it can be concluded that Flow Cylinder test is more sensitive than the Flow Table test. Test on sand also support this observation. Flow Table test only showed a slight variation between the flowbility of sand and cement, but the Flow cylinder test showed different results. While the sand flowability was over 200% the cement flowability was reduced to 60% (as compared to the Flow Table test).

Wet Materials: Flowability of wet samples were investigated. Water was added to the binders to achieve a flowbility of 100% using the Flow Table method. Fly ash-F, fly ash-C and cement required 22, 30 and 43 percent of water respectively. Cement required almost double the amount of water that was required by fly ash-F to achieve similar flowability (Flow Table Method). These water-binder mixtures had different responses with the Flow Cylinder method. While the cement had similar flowability (100%) with both the test methods, responses with fly ashes were different.

Flowable Fill: Flowability of Mix -1 and Mix-2 are summarized in Table 1. Slump for the two mixtures were 200 mm. The bleeding of the two mixtures are also summarized in Table 1.

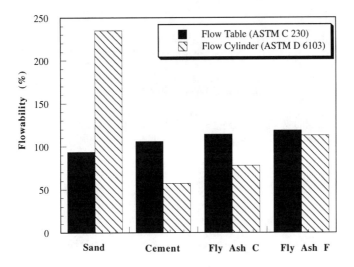

Figure 1. Flowability of dry materials.

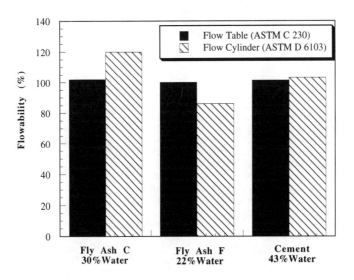

Figure 2. Flowability of wet binder materials.

(b) Setting Time

Binders: Typical needle penetration-time relationships for cement (43% water content), fly ashes (Class F (22% water content) & C (30% water content)) are shown in Fig. 3. Fly ash-C showed rapid setting as compared to the other two binders. Cement had an initial and final setting of 3 and 6 hours respectively. Fly ash-F had an initial and final setting times of 10 and 17 hours respectively.

Flowable Fill: Changes in penetration resistance of Mix-1 and Mix-2 flowable fills with varying amounts of cement are shown in Figs. 4 and 5 respectively. The theoretical water requirement for the cement and fly ash-F in Mix-1 were 1.3 and 2.6 % respectively. Hence the excess water available for making the Mix-1 to flow was 6.5%. The theoretical water requirement for the cement and fly ash in Mix-2 were 0.65 and 3.0 % respectively. Hence the excess water available for making the Mix-2 to flow was 6.75%.

Mix-1: The penetration resistance after 24 hours was 2 MPa (300 psi). The initial setting time was 35 hours (based on 3.4 MPa (500 psi) resistance). The penetration resistance was 13.8 MPa (2000 psi) after 100 hours. According to ASTM C 403, the final setting time is when the penetration resistance is 27.6 MPa (4000 psi).

Mix-2: The penetration resistance after 24 hours was 1.4 MPa (200 psi). The initial setting time was 70 hours (based on 3.4 MPa (500 psi) resistance). The penetration resistance was 4.1 MPa (600 psi) after 100 hours.

Hence doubling the cement content increased the penetration resistance by 50% in 24 hours. It also increased the penetration resistance by over three fold in 100 hours.

(c) Strengths

(i) Binders

Compression: Compressive strength of the binders increased with curing at different rates for all three binders. The greatest increase in strength was observed within the first three days. Variation of binder strength with curing time up to seven days are shown in Fig. 6. The percentage increases in strength from the third to the seventh day were 30, 19 and 14 % for fly ash-C, cement and fly ash-F respectively. Largest increase in strength was observed in cement. The seventh day compressive strength of fly ash-C was one-third of the strength of cement and the strength of fly ash-F was one-thirtieth of fly ash-C.

Splitting Tension: Tensile strengths of the binders increased with curing at different rates for all three binders. The greatest increase in strength was observed within the first three days. Variation of binder strength with curing time up to seven days are shown in Fig. 7. The percentage increases in strength for fly ash-C, cement and fly ash-F from the third to the seventh day were 71, 21 and 3 % respectively. The seventh day compressive strength of fly ash-C was half of the strength of cement and the strength of fly ash-F was one-fortieth of fly ash-C. The relationship between compression strength (σ_c) and tensile strength (σ_t) for all three binders are linear (Fig. 8) and can be represented as follows:

$$\sigma_t = m \, \sigma_c \tag{1}$$

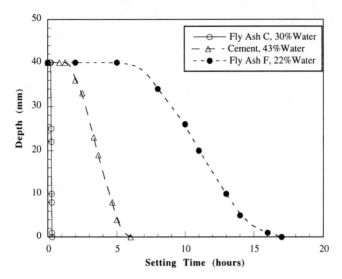

Figure 3. Setting times (Vicat needle method) of binder materials

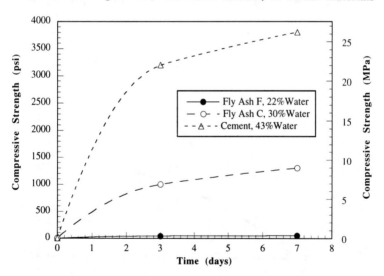

Figure 6. Variation of compressive strength with curing time for cement, fly ash-C and fly ash-F.

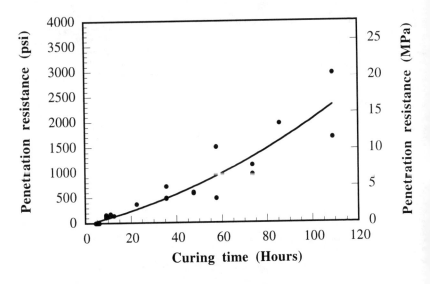

Figure 4. Penetration resistance versus curing time for Mix-1.

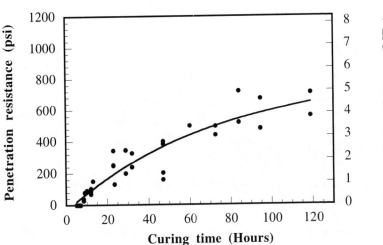

Figure 5. Penetration resistance versus curing time for Mix-2.

Where m is the strength parameter (slope of the line). The results are summarized in Table 2.

Table 2. Strength Parameter (m) for Binders and Flowable Fills

Materials	Cement	Fly ash -C	Fly ash-F	Mix-1	Mix-2
Parameter m	0.11	0.16	0.12	0.14	0.12
Coefficient of Correlation	0.99	0.99	0.99	0.99	0.99

Flowable Fill

Mix-1

Compression: Variation of compressive strength with curing time is shown in Fig. 9. The third day strength was 0.7 MPa (100 psi). The increase in strength from the third to the seventh day was 100 % and form 7th to 28th day was 290%. The average 28th day compressive strength was 5.7 MPa (830 psi).

Splitting Tension: Variation of splitting tensile strength with curing time is shown in Fig. 9. The third day strength was 0.1 MPa (15 psi).The percentage increase in strength from the third to the seventh day was 150% and form 7 to 28 day was 220%. The relationship between tensile and compressive strength was linear (Fig. 10) and the result is summarized in Table 2.

Mix-2

Compression: Variation of compressive strength with curing time is shown in Fig. 11. The third day strength was 0.4 MPa (54 psi). The increase in strength from the third to the seventh day was 83% and form 7 to 28 day was 140%. Increasing the cement content in the mix from 1.5 to 3% increased the 28th day strength from 1.6 MPa (240 psi) to 5.7 MPa (830 psi), 350% increase.

Splitting Tension: Variation of splitting tensile strength with curing time is shown in Fig. 11. The third day strength was 0.05 MPa (7 psi). The increase in strength from the third to the seventh day was 70 % and form 7 to 28 day was 130. Increasing the cement content in the mix from 1.5 to 3% increased the 28th day strength from 0.19 MPa (28 psi) to 0.85 MPa (123 psi), 440% increase. The relationship between tensile and compressive strength was linear (Fig. 12) and the result is summarized in Table 2.

Conclusions

Behavior of binders and two flowable fill mixtures with fly ash (Class-F) were characterized based on their working and mechanical properties. Based on the experimental results the following conclusions are advanced:

1. **Flowability**: Flow Table and Flow cylinder results for the dry materials were not in agreement. Flow cylinder showed the largest variation between the materials investigated. Difference between the two methods are reduced

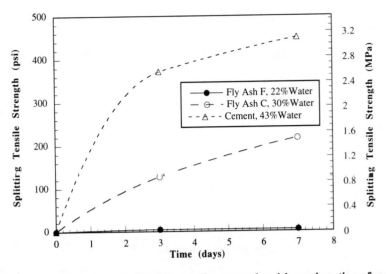

Figure 7. Variation of splitting tensile strength with curing time for cement, fly ash-C and fly ash-F.

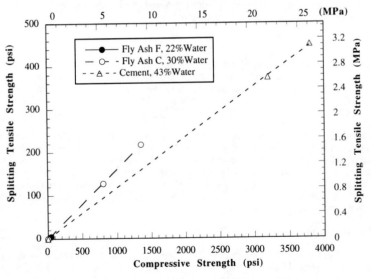

Figure 8. Relationship between compressive and splitting tensile strengths of the binder materials

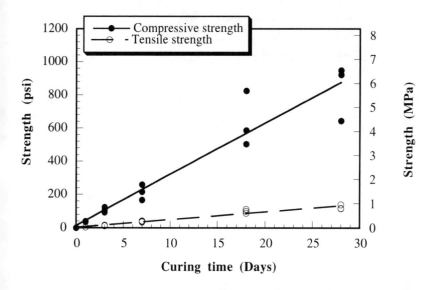

Figure 9. Variation of compressive and tensile strengths with curing time for Mix-1.

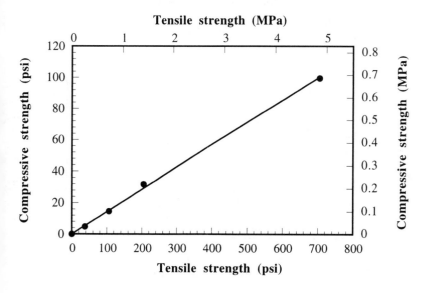

Figure 10. Relationship between compressive and splitting tensile strengths of Mix-1

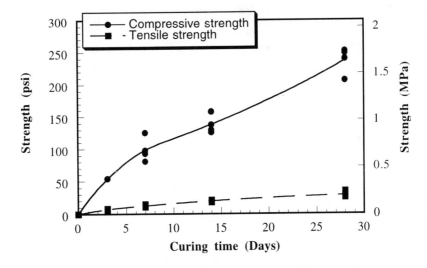

Figure 11. Variation of compressive and tensile strengths with
curing time for Mix-2.

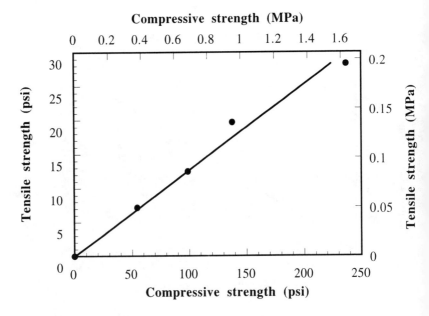

Figure 12. Relationship between compressive and splitting tensile
strengths of Mix-2

with the flowable wet binders. Cement required almost double the water that was required by Fly ash-F for similar flowability.

2. **Setting time**: Of the binders, fly ash (Class C) with 30% water had the shortest setting time as compared to cement (43%) and fly ash-F (22%). The penetration test can be used to determine the setting time and initial strength development of the flowable fill. Doubling the cement content penetration resistance by 50% in the first 24 hours.

3. **Strength**: The compressive and splitting strengths of the binders and flowable fills increased with curing time at varying rates. Linear relationship between the tensile and compressive strength were observed for the binders and flowable fills. The tensile-to-compressive strength ratios varied from 0.11 to 0.16. Doubling the cement content from 1.5 to 3% increased the 28-day compressive and tensile strengths of the flowable fill by 350 and 440% respectively.

Acknowledgment

This work was supported by the Center for Innovative Grouting Materials and Technology (CIGMAT) under grants from the National Science Foundation (CMS-9526094) and the City of Houston.

References

[1] ACI Manual of Concrete Practice Part 1(1997), Controlled Low Strength Materials (CLSM) Reported by ACI Committee 229 (ACI 229R-94).

[2] Adaska, W. S. (Editor) (1994), Controlled Low Strength Materials, SP-150, ACI, 113 p.

[3] Annual Book of ASTM Standards (1997), Volume 04.02 Concrete and Mineral Aggregates, ASTM, Philadelphia, PA.

[4] Ayers, M. E., Wong, S. Z. and Zaman, W. (1994), " Optimization of Flowable Fill Mix Proportions," SP-150, ACI, pp. 15-38.

[5] Diamond, S., "On the Glass Present in Low-Calcium and in High-Calcium Fly Ash," Cement and Concrete Research, Vol. 13, No. 4, July 1983, pp. 459-464.

[6] Hill, J. C. and Sommers, J. (1997), "Production and Marketing of Flowable fill Utilizing Coal Combustion By-Product," Proceedings , American Coal Ash, 12th International Symposium, Vol. 2, pp. 38-1 through 38-13.

[7] Hoff, G. C. (1971), "New Applications for Low-Density Concrete," ACI Publication SP 29 on Lightweight Concrete, pp. 181-186.

[8] Huang, W. H.(1997), "Properties of Cement-Fly ash Grout Admixed with Bentonite, Silica fume, or Organic Fiber," Cement and Concrete Research, Vol. 27, No. 3, pp. 395-405.

[9]   Janardhanam, R., Burns, F. and Peindl, R. D. (1992), "Mix Design for Flowable Fly-ash Backfill Materials," Journal of Materials in Civil Engineering, ASCE, Vol. 4, No. 3, pp. 252-263.

[10] Landwermeyer, J. S. and Rice, E. K. (1997), "Comparing Quick-Set and Regular CLSM," Concrete International, ACI, Vol. 19, No. 5, pp. 34-39.

[11] Naik, T. R. and Singh, S. S. (1997), "Flowable Slurry Containing Foundry Sands," Journal of Materials in Civil Engineering, ASCE, Vol. 9, No. 2, pp. 93-102.

[12] Ramme, B., and Naik T. R. (1998), "Controlled Low-Strength Materials (CLSM) State-of-the -Art , " Presented ACI Spring 1998 Convention, March 22-27, Houston, Texas.

[13] Ramme, B., Naik T. R. and Kolbeck, H. J. (1995), "Construction Experience with CLSM Fly ash Slurry for Underground Facilities," ACI Special Publication 153, pp. 403-416.

[14] Roy, D. M., Luke, K., and Diamond, S., "Characterization of Fly ash and Its Reactions in Concrete," Proceedings, Materials Research Society, Pittsburgh, 1984, pp. 124-149.

[15] Smith, A., (1991), "Controlled Low-Strength Materials," Concrete Construction, pp. 389- 398.

The Safe Disposal of Fly Ash in Pavement or Earth Structures Not Requiring High Strength Materials

Vivek Tandon, A.M. ASCE [1] and Miguel Picornell, M. ASCE [2]

Abstract

Highway agencies throughout the country face the challenge of safe disposal of various waste products in pavements and earth structures. Some of these waste products can be hazardous to the environment as suggested by the EPA. One such product is fly ash, which is produced during coal combustion in combination with bottom ash. More than 70 million tons of fly ash are produced in the USA each year. Safe disposal of such an enormous quantity of material is critical. One possible destination is the 2.3 million miles of highways. Various researchers and highway agencies have used fly ash but in small quantities. More uses are needed.

The present study was undertaken to evaluate the use of fly ash in earth structures not requiring high strength such as vertical moisture barriers, shoulders, or as a backfill material. Fly ash was used in combination with cement and sand to form a fly ash-cement-sand mortar. The results of the study are reported in this paper.

Introduction

The U.S. Portland cement industry, the electrical utilities, the EPA, and standard-setting agencies such as ASTM are attempting to deal with the technically challenging problems posed by the need for energy conservation, environmental protection, and economy. An important aspect of these problems concerns the maximum utilization of waste materials. The amount of fly ash being generated by coal-fired electric generating stations is increasing rapidly to levels approximating

---

[1] Research Engineer, [2] Associate Professor, Department of Civil Engineering, UTEP, El Paso, Texas 79968. Phone: (915) 747-6924, Email: vivek@utep.edu

the volume of Portland cement being produced. However, less than 20% of the ash is used, creating a substantial waste-disposal and environmental problem (Helmuth, 1987). Safe disposal of such an enormous quantity of material is critical. The 2.3 million miles of highways is a possible destination to reuse large quantities of these materials.

Various researchers and highway agencies have used fly ash but in small quantities. The most common use of fly ash is as a replacement of Portland cement in Portland cement concrete. Although this use allows the safe disposal of fly ash, the quantity of fly ash disposed is small in comparison to fly ash being produced. To dispose of larger quantities safely, new means of fly ash disposal are needed. One means can be the use of fly ash in lightly loaded areas of transportation facilities such as shoulders, as backfill in embankments, etc.

Fly ash contains heavy metals such as Ba^{2+}, Cu^{2+}, Cd^+ etc. These heavy metals are hazardous to the environment in concentrations above certain limits, as set by the EPA. Thus, evaluating whether fly ash can be used in an environmentally safe manner is necessary. Hence, one major consideration should be stabilization of the heavy metals. To dispose of large quantities safely, fly ash can be used in combination with cement and sand to form a fly ash-cement-sand mortar. The use of cement allows the stabilization of heavy metals and simultaneously provides some strength to the material.

Objective and Scope of Work

The overall objective of this study was to evaluate the economical use of fly ash in structures not requiring high strength material. The fly ash-cement-sand mortar was evaluated in terms of strength, permeability, shrinkage potential, and mobility of heavy metals present in the fly ash.

An important consideration was to document the effect of Portland cement on the mobility of the heavy metals present in fly ash. Barium was selected as the metal to document this effect. This selection was based on the relative abundance of this element in fly ash. Nevertheless, it is important to realize that the mobility of other heavy metals present in fly ash might be influenced differently than Barium by the addition of Portland cement.

The specimens used in this study were conditioned in a water bath at 61 °C (142 °F) for seven days, which is equivalent to 28 days of conditioning at room temperature. The main purpose was to reduce the time for specimen preparation. Fly ash "Type C" was used as a replacement of cement as well as sand in a fly ash-cement-sand mortar. Three cement contents were used in this analysis 3, 5, and 7% (by weight). A minimum of 3% was selected because the mortar was not workable for less than 3% cement. A maximum of 7% of cement was selected based on cost considerations. Three fly ash to sand ratios: 25/75, 50/50, and 75/25 were selected to cover the possible range. Two levels of water, i.e., water-cement ratios 3 and 3.25 were selected because below 3, the mortar was not workable and above 3.25, the mortar was too fluid to prepare specimens. Only two water-cement ratios of 3 and 3.25 were used because of the time constraints imposed by the project. Eighteen specimens were prepared for each combination of test variables.

The following tests were performed to evaluate the fly ash-cement-sand mortar: 1) compressive strength test, 2) permeability test, 3) length changes test, 4) acid digestion of sludge's and atomic absorption test, and 5) leaching tests. The test methodologies are described in the following section.

Laboratory Test Methods

The laboratory tests used in the evaluation of fly ash-cement-sand mortar are described in this section. All of the selected test methods were standard test procedures. If any of the steps of the standard test procedures were changed, these changes are described along with each test procedure.

Compressive Strength Test

The compressive strength tests were performed following ASTM standard C-109-84. In total, 18 specimens were prepared for compressive strength tests. All the standard test procedures were followed with the following two exceptions. As per the ASTM standard, the specimens should be prepared in a cubic mold of 5 cm (2 in.). However, the specimens in the present study were prepared in cylindrical molds of 7.6 cm (3 in.) diameter and 15 cm (6 in.) high. The other difference was the curing process. The specimen was cured in a water bath at 61 °C (142 ° F) for 7 days rather than being cured in humidity chamber for 28 days.

Permeability Test

The permeability test was performed on a specimen as per Corps of Engineers Standard CRD-C 163 (1992). This test was performed to find the hydraulic conductivity and porosity of the mortar. The permeability test was performed following this standard with one exception. The deviation was the application of driving pressure. The driving pressure was applied directly to the water rather than using a gas water accumulator. Since the tests were only performed for short duration's, the effect of applying the driving pressure directly to the water was expected not to be significant.

Length Change Test

The length change tests were performed according to ASTM standard C-157-89. This test was performed to observe the change in length of the fly ash-sand-cement mortar during the curing period. The test results obtained from this test reflect how much expansion or contraction a mortar would have, after placing the mortar as backfill. These specimens were also cured in a water bath for 7 days at 61 °C (142 ° F) for faster curing. Another deviation from the standard test procedure was that the water bath consisted of regular tap water rather than lime saturated water.

Acid Digestion of Sludges and Atomic Absorption Test

A literature review indicated that fly ash contains heavy metals. Gupta and Ray (1993) indicated that fly ash contains heavy metal elements like Barium, Lead, Arsenic, Mercury etc. in detectable quantities. Although the quantity of these elements is quite small, the elements are considered hazardous (EPA, 1989a). The sludge samples for atomic absorption tests were prepared using the acid digestion of sludges procedure (EPA, 1989b). The atomic absorption test was performed to determine the concentration of barium in the fly ash. The barium element was selected, since existing literature indicates that it is one of the most abundant. The acid digestion of sludges was performed as per EPA Test method 3050 (EPA, 1986b) and the atomic absorption test was performed as per the EPA method 7080 (EPA 1986c).

Extraction Procedure Toxicity Test Method

This test was performed as per the EPA standard test method 1310A (EPA, 1986a). This test was performed to evaluate the potential for leaching of the heavy metals from the mortar. The test was performed exactly as outlined in the test procedure.

Test Results and Discussions

Compressive Strength Test Results

The results of the compressive strength tests are presented in Table 1. Minimum compressive strength was observed in specimens with 3% cement, 75/25 fly ash-sand ratio for both water-cement ratios. All the specimens prepared using 3% cement were difficult to work with and were breaking apart while being taken out of the mold or while moving them to perform the compressive strength test.

Table 1. Summary of Compressive Strength Test Results

Water-Cement Ratio	Fly ash-Sand Ratio	Compressive Strength (MPa)		
		3% Cement	5% Cement	7% Cement
3.00	25/75	5.8	11.4	3.8
	50/50	0.9	13.0	16.8
	75/25	0.3	7.0	18.9
3.25	25/75	4.4	9.6	7.9
	50/50	1.5	9.6	9.7
	75/25	0.3	3.7	9.6

Most of the specimens prepared using 5 or 7 % cement showed good compressive strength capabilities except for two specimens. Both specimens were prepared with 7% cement and 25/75 fly ash-sand ratio, and for water-cement ratios of 3.00 and 3.25. The compressive strength for these two specimens was lower than for the specimens prepared with 5% cement. Usually the compressive strength of mortar specimens increases for increasing percentage of cement. However, in the present case it decreased. This behavior did not seem reasonable, so it was decided to prepare two additional specimens with 7% cement and 25/75 fly ash-sand ratio, and for water-cement ratios of 3.00 and 3.25. The test results on the new specimens showed results similar to the initial specimens, i.e., 7% cement specimens had lower strength as compared to 5% cement.

A rational explanation is not apparent at this time and more research would be needed to clarify this point. The specimens that showed higher compressive strength were prepared with 7% cement, 50/50 fly ash-sand ratio or higher and a water-cement ratio of 3, as shown in Table 1. If the proportions needed to be chosen solely based on workability and compressive strength, the recommended proportions would be 7% cement, 75/25 fly ash-sand ratio, and a water-cement ratio of 3. This combination would be good, from a recycling point of view, since the mortar includes a large amount of fly ash.

Permeability Test Results

A sample of the data obtained from a permeability test is shown in Table 2 and Figure 1. Table 2 shows the permeability test performed on a specimen prepared with 3% cement, 75/25 fly

Table 2. Permeability Test Results of Specimen Consisting of 3% Cement 75/25 Fly ash-Sand Ratio with Water-Cement Ratio of 3.00

Specimen Id. 18   Date of Experiment: 8/17/95 File Name: SPCMN18
Cement: 3%   Fly ash 72.78% Sand 24.25% Water/Cement Ratio: 3.00
Average Diameter of Specimen (cm):   (7.2567+7.2644+7.2898)/3= 7.27
Average Height of Specimen (cm):   (7.2898+7.2898+7.2898)/3= 7.29
Cross-sectional Area of Specimen: 41.51 cm^2
Viscosity of Water: 9.33E-04   Pa-S   Density of Water: 1.00   gm/cm^3
Confining Pressure: 344.74   KPa   Driving Pressure: 275.79 KPa
Atm. Pressure :   101.33   KPa
Saturated Surface Dry Weight of Specimen: 580.80 gm
Dry Weight of Specimen: 480.80 gm
Submerged Unit Weight of Specimen: 286.20 gm

Reading No.	Time Elapsed (sec)	Incremental Time (sec)	Total Volume	Incremental Vol. Collected (ml)	Intrinsic Permeability (1E^{-9} cm^2)	Hydraulic Conductivity (1E^{-5} cm/sec)
1	0	0	0	0		
2	5	5	9.5	9.5	1.78	1.75
3	10	5	17.5	8.0	1.50	1.47
4	15	5	25.5	8.0	1.50	1.47
5	20	5	33.5	8.0	1.50	1.47
6	25	5	41.4	7.9	1.48	1.45
7	30	5	49.3	7.9	1.48	1.45
8	35	5	57.3	8.0	1.50	1.47
9	40	5	65.2	7.9	1.48	1.45
10	45	5	73.2	8.0	1.50	1.47

ash-sand ratio, and water-cement ratio of 3.0. The permeability tests were continued until the ratio of the increments of volume and time becomes linear. Figure 2 shows that at the beginning the ratio was nonlinear, but after 10 seconds of testing, the specimen achieved a steady state condition.

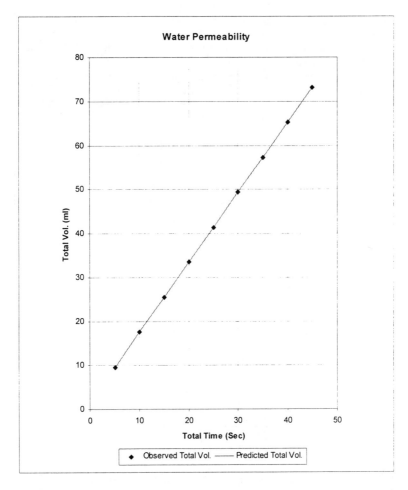

Figure 1. Hydraulic Conductivity of Specimen Consisting of 3% Cement, 75/25 Fly ash-Sand Ratio with Water-Cement Ratio of 3.00

The hydraulic conductivity of all the tests are summarized in Table 3. The hydraulic conductivity ranged from 14.7 to 0.037 10^{-7} cm/sec. The minimum hydraulic conductivity was observed in the specimens prepared with 5% cement, 75/25 fly ash-sand ratio, and water-cement ratio of 3.00. The maximum hydraulic conductivity was observed in the specimens prepared with 3% cement, 75/25 fly ash-sand ratio, and water-cement ratio of 3.00. The specimens prepared with 5% or 7% cement showed lower hydraulic conductivities especially with water-cement ratios of 3.00.

The porosity of fly ash-sand-cement mortar can also be obtained from the specimens prepared for permeability. After measuring permeability, the specimens were taken out and the saturated surface dry weight was measured. Then each specimen was kept in oven for 24 hours to measure the dry weight of the specimen. The ratio of volume of water to the total volume is the porosity in %. The porosity obtained from the tests is given in Table 4. The porosity of all the proportions ranged from 16.8% to 38.1%. The specimens prepared with 5% cement, 25/75 fly ash-sand ratio and water-cement ratio of 3.25 exhibited the lowest porosity. The highest porosity was obtained for 3% cement, 50/50 fly ash-sand ratio and a water-cement ratio of 3.00.

Table 3. Summary of Permeability Test Results

Water-Cement Ratio	Fly ash-Sand Ratio	Hydraulic Conductivity (10^{-7} cm/sec)		
		3% Cement	5% Cement	7% Cement
3	25/75	11.3	0.05	0.80
	50/50	3.60	0.07	0.06
	75/25	14.7	0.037	0.85
3.25	25/75	0.72	0.23	0.43
	50/50	7.90	0.04	0.20
	75/25	1.57	0.10	0.09

Table 4. Summary of Estimated Porosity of Fly Ash-Sand-Cement Mortar

Water-Cement Ratio	Fly ash-Sand Ratio	Porosity (%)		
		3% Cement	5% Cement	7% Cement
3	25/75	27.5	30.1	22.9
	50/50	38.4	32.6	21.9
	75/25	34.9	30.3	23.6
3.25	25/75	18.9	16.8	18.0
	50/50	23.9	21.2	19.1
	75/25	25.2	31.3	33.1

Length Change Test Results

The length change test was performed on 18 specimens and the results obtained from these tests are shown in Table 5. The results obtained from the tests indicate change in length of specimens during the curing period. The positive sign indicates that the specimen expanded (%) and the negative sign indicates that the specimen shrank (%). The maximum expansion of 0.12% was observed in the specimen prepared with 3% cement, 75/25 fly ash-sand ratio, and a water-cement ratio of 3.

The largest shrinkage of 0.012% was observed in the specimen prepared with 7% cement, 25/75 fly ash-sand ratio, and a water-cement ratio of 3.25. The test results showed that the length change might be affected by the water-cement ratio. In this manner, five specimens shrank for a water-cement ratio of 3.25 while only two specimens shrank for a water-cement ratio of 3.0. However, more testing may be needed to clarify the effect that the water bath had on the curing of specimens in terms of the length change. It can be concluded from the above test results that the proportions do not affect the length change.

Table 6. Summary of Length Change Test Results

Water-Cement Ratio	Cement (%)	Fly ash-Sand Ratio	Initial Length[+], L_i (cm)	Final Length[*], L_x (cm)	Change in Length[++], ΔL (%)
3	3	25/75	0.025	0.017	-0.033
		50/50	-0.010	-0.015	-0.020
		75/25	-0.041	0.010	0.120
	5	25/75	-0.005	0.018	0.09
		50/50	0.028	0.049	0.08
		75/25	0.020	0.024	0.016
	7	25/75	-0.046	-0.018	0.110
		50/50	-0.005	-0.025	0.010
		75/25	-0.006	0.000	0.025
3.25	3	25/75	0.015	0.009	-0.024
		50/50	0.015	0.011	-0.017
		75/25	-0.010	-0.009	0.005
	5	25/75	-0.005	-0.066	-0.060
		50/50	0.001	0.008	0.028
		75/25	0.027	0.050	0.090
	7	25/75	0.015	0.012	-0.012
		50/50	-0.008	-0.012	-0.018
		75/25	-0.008	0.001	0.025

[+]L_i= Difference Between the Length of the Specimen and Reference Bar Before Curing

[*]L_x= Difference Between the Length of the Specimen and Reference Bar After 28 Days Curing

[++]$\Delta L = (L_x - L_i)*100/G$

G= Length of the Reference Bar, 254 mm (10 in.)

Barium Concentrations in Fly ash

The atomic absorption test was performed on the eight samples, prepared using acid digestion of sludges method, of "Type C" fly ash. All the samples were collected and tested randomly. The purpose of this test was to find the actual quantity of Barium present in the fly ash. All the samples were digested for a fixed period and then the concentration of barium in the fluid was measured using an atomic absorption spectrometer. The results are given in Table 6. The amount of Barium varied from 3300 mg/Kg to 1465 mg/Kg with an average of 2313 mg/Kg. The average value of 2313 mg/Kg was used to calculate percentages of barium leached out of specimens.

Table 6. Atomic Absorption Test Results of Fly ash

Sample Number	Absorbency		Average Absorbency	Concentration (ppm)	Ba(mg)/kg of Fly ash
	Reading 1	Reading 2			
1	.023	0.023	0.0225	47.50	2375.00
2	.027	0.027	0.0270	66.00	3300.00
3	.018	0.019	0.0185	35.66	1783.00
4	.024	0.021	0.0225	45.80	2290.00
5	.019	0.019	0.0190	37.00	1850.00
6	.022	0.023	0.0225	45.80	2290.00
7	.026	0.027	0.0265	63.00	3150.00
8	.016	0.016	0.0160	29.30	1465.00
Average Concentration of Ba (mg)/Kg of Fly ash					2313

Extraction Procedure

The leaching test was performed on replicate specimens, i.e., 36 specimens. Table 7 shows the amount of barium in the extracted solution and the percentage of barium leached out. The amount of barium leaching out was higher in all the specimens prepared with 25/75 fly ash-sand ratio. The results indicate that as the fly ash-sand ratio increases, the amount of fly ash leaching out decreases. As per EPA Federal Register (1990) recommendations, the amount of heavy metal, in this case Barium, in the extract should be below 100 mg/Liter and all the specimens are well below the limit; minimum of 2.88 mg/L and a maximum of 14.1 mg/L, specified by the EPA. However, the amount of Barium leaching out from the specimens is lower in the specimens with 50/50 and 75/25 fly ash-sand ratio as compared to 25/75 fly ash-sand ratio. Based on results obtained from extraction test specimens prepared with 75/25 fly ash-sand ratio, a water-cement ratio of 3.00 for both 5% and 7% cement can be used safely.

These tests were performed to evaluate the mobility of Barium. Extrapolating this behavior to other heavy metals is not possible. Therefore, a more comprehensive testing program should be implemented before the stabilized fly ash can be safely used as a trench backfill material.

Conclusions

Based on the test results, it can be concluded that using fly ash-sand-cement mortar in structures not requiring high strength is feasible. However, more testing is required to ensure safe environmental usage of fly ash-sand-cement mortar. The test program described showed that 7% cement, 75/25 fly ash-sand ratio, with a water-cement ratio of 3.00 is probably the best compromise.

Based on the unconfined compressive strength, the maximum strength of 18.9 MPa was obtained for the 7% cement, 75/25 fly ash-sand ratio, and a water-cement ratio of 3.00. These specimens showed a hydraulic conductivity of 0.85×10^{-7} cm/sec. This hydraulic conductivity is not the lowest by an order of magnitude; however, the lowest hydraulic conductivity results in a reduction of compressive strength to one third of the strength of the proposed mix.

The results of the leaching tests clearly indicate that the larger cement content specimens resulted in lower percentages of barium leaching out of the specimens. Thus, from this point of view of immobilizing the heavy metals, the proposed mix provides one of the best results obtained in the test program. This consideration will probably hold for other heavy metals. However, independent confirmation for other specimens would be wise before any disposal in the field.

Table 7. Barium Leaching from Fly ash-Sand-Cement Mortar

Water-Cement Ratio	Cement (%)	Fly ash-Sand Ratio	Absorbency Reading Specimen		Ba Conc. in Extract (mg/L) Specimen		Avg. Amount of Ba Leached out (mg)	Barium Amount in the Specimen (mg)	Barium Amount Leached out (%)
			1	2	1	2			
3.00	3	25/75	0.013	0.011	9.12	7.75	16.53	56.13	29.45
		50/50	0.009	0.010	6.37	7.12	13.22	122.27	10.81
		75/25	0.013	0.007	9.12	4.75	13.59	168.40	8.07
	5	25/75	0.013	0.013	9.12	9.12	17.88	55.07	32.46
		50/50	0.009	0.008	6.37	5.62	11.75	110.14	10.67
		75/25	0.021	0.007	14.1	4.75	18.49	165.20	11.19
	7	25/75	0.020	0.008	13.5	5.62	18.74	54.04	34.67
		50/50	0.008	0.007	5.62	4.75	10.16	108.08	9.40
		75/25	0.013	0.005	9.12	2.88	11.76	162.12	7.25
3.25	3	25/75	0.014	0.012	9.75	8.4	17.79	56.13	31.69
		50/50	0.012	0.007	8.40	4.75	12.89	122.27	10.54
		75/25	0.006	0.011	3.87	7.75	11.39	168.40	6.76
	5	25/75	0.014	0.010	9.75	7.12	16.53	55.07	30.02
		50/50	0.010	0.006	7.12	3.87	10.77	110.14	9.78
		75/25	0.009	0.006	6.37	3.87	10.04	165.20	6.07
	7	25/75	0.011	0.009	7.75	6.37	13.84	54.04	25.61
		50/50	0.008	0.006	5.62	3.87	9.30	108.08	8.60
		75/25	0.009	0.005	6.37	2.88	9.07	162.12	5.59

Acknowledgments

This research project was funded by the Texas Department of Transportation and the authors gratefully acknowledged this support. The support and guidance of Elias H. Rmeili, the project director is sincerely appreciated.

References

1. EPA (1990), Federal Register 55, 11862 (3/29/90), RCRA Toxicity Characteristic ,Final Rule.
2. EPA (1989), Stabilization/Solidification of CERCLA and RCRA Wastes, United States Environmental Protection Agency, EPA/625/6-89/022, May.
3. EPA Test Method 1310A (1986a), Test Methods For Evaluating Solid Waste, United States Environmental Protection Agency, Volume 1, Section C, pages 1310A/1 1310A/18, September.
4. EPA Test Method 3050 (1986b), Test Methods For Evaluating Solid Waste, United States Environmental Protection Agency, Volume 1, Section A, pages 3050/1-3050/3, September.
5. EPA Test Method 7080 (1986c), Test Methods For Evaluating Solid Waste, United States Environmental Protection Agency, Volume 1, Section A, pages 7080/1-708 0/3, September.
6. Gupta, P.C. and Ray, S.C., (1993), "Commercialisation of Fly ash," The Indian Concrete Journal, Vol 67, No. 11, pages 554-558,November.
7. Helmuth, R., (1987)," Fly Ash in Cement and Concrete," Portland Cement Association, Research and Development Laboratories, Skokie, Illinois, pp. i.
8. Standard CRD-C 163 (1992), "Test Method for Water Permeability of Concrete Using Triaxial Cell," Corps of Engineers Standard Test Procedure, September.

CRUSHED HYDRATED FLY ASH AS A CONSTRUCTION AGGREGATE

Sanjaya P. Senadheera[1], Associate Member, ASCE
Priyantha W. Jayawickrama[2], Member, ASCE and Ashek S.M. Rana[3]

ABSTRACT: Hydrated fly ash (HFA) is being produced in Texas by first adding water to Class C fly ash and allowing it to hydrate in curing pits for several weeks. HFA typically attain unconfined compressive strengths as high as 15 MPa. Once the HFA gains an acceptable level of strength, it is crushed using specialized equipment to produce an aggregate material with characteristics very similar to those of conventional granular materials such as crushed limestone. In certain regions of Texas, where quality aggregate base materials are in short supply, this synthetic aggregate material has been used in the construction of road bases on an experimental basis. Experience from the use of HFA ash in these different regions has been quite variable. This paper provides an overview of HFA as a construction aggregate, and documents the experience from its use as a road base material.

INTRODUCTION

Fly ash is an inorganic powdery by-product of coal burning. According to the latest numbers published by the Electric Power Research Institute (EPRI), the total coal ash production in US is 72 Mg/year. Out of this, approximately 74 percent is fly ash (Bloomquist et al., 1993). Bloomquist et al. (1993) also reported that, in 1986, approximately 18 percent of this fly ash were put into effective use and the remainder dumped in landfills. At the current production levels, and for a 20 percent utilization level, fly ash dumped in landfills occupies approximately 7,600 acre-meters of landfill volume every year.

---

[1] Research Asst. Professor, Dept. of Civil Engineering, Texas Tech University, Lubbock, TX 79409
[2] Assoc. Professor, Dept. of Civil Engineering, Texas Tech University, Lubbock, TX 79409
[3] Doctoral Candidate, Dept. of Civil Engineering, Texas Tech University, Lubbock, TX 79409

Chemical composition in fly ash from different sources is quite variable. Its primary constituents include silica (SiO_2), alumina (Al_2O_3) and various other oxides and alkalis. There is also considerable variability in the quality and reactivity of fly ash obtained from different sources. Such variability may be due to differences in coal composition from different sources, coal burning processes, and the fly ash collection method in the power plant. ASTM C-618 classifies fly ash as either Type C or Type F based on the chemical composition. Type C fly ash has a higher proportion of calcium compounds.

Hydrated Fly Ash

Type C fly ash used in this research, when mixed with water, formed a hard, homogeneous rock-like material which is referred to as hydrated fly ash (HFA). This self-hardening property of Type C fly ash in the presence of water, and the resulting significantly high strengths have drawn the interest of civil engineers.

In Texas, HFA is currently being produced by dumping Type C fly ash into curing pits in layers approximately 25-30 cm thick. Water is typically sprayed to each layer of fly ash and it is allowed to hydrate for 2 to 3 weeks. It was observed that the amount of water used varied between different HFA production facilities. It was also observed that there is a lack of adequate control of the amount of water added for hydration. One of the objectives of this research is to study the influence of curing water content on the properties of HFA. Once HFA attains acceptable strength levels, the homogeneous mass of hydrated fly ash is crushed inside the curing pit using specialized equipment to produce an aggregate material that is similar to conventional granular materials such as crushed limestone or gravel.

The major pozzolanic reactions applicable to the hydration of fly ash were presented by Pollard et al. (1992).

$Ca(OH)_2 + SiO_2 + H_2O \rightarrow (CaO)_x(SiO_2)_y(H_2O)_z$
calcium silicate hydrates

$Ca(OH)_2 + Al_2O_3 + H_2O \rightarrow (CaO)_x(Al_2O_3)_y(H_2O)_z$
calcium aluminate hydrates

$Ca(OH)_2 + Al_2O_3 + SiO_2 + H_2O \rightarrow (CaO)_x(Al_2O_3)_y(SiO_2)_z(H_2O)$
calcium aluminate silicate hydrate

$Ca(OH)_2 + Al_2O_3 + SO_3 + H_2O \rightarrow (CaO)_x(Al_2O_3)_y(CaSO_3)_z(H_2O)$
calcium aluminate calcium silicate hydrates

The coefficients w, x, y, and z can assume a range of values. These hydration products are generally poorly crystallized and have very high surface area. They are quite similar to those formed in the hydration of Portland cement. Therefore it can be postulated that hydrated calcium silicates and/or calcium aluminates are mainly responsible for the strength development during hydration of both Portland cement and fly ash (Sanders et al, 1995).

Although not published in technical literature, HFA has been used in Texas for quite some time in oil field haul roads, city and county streets, parking lots and driveways. In addition, the Texas Department of Transportation (TxDOT) has used this material in base course construction on several experimental construction projects (Grant 1995). The primary objectives of this research were the evaluation of hydrated fly ash as an alternative construction aggregate and to develop specifications for its use. The work involved characterization of HFA in the laboratory and the evaluation of HFA performance in projects undertaken by TxDOT.

HYDRATED FLY ASH: SOURCES, CHEMICAL COMPOSITION AND PHYSICAL PROPERTIES

HFA used in TxDOT experimental pavement construction projects came from three different sources. Table 1 identifies the four TxDOT districts where the experimental projects were completed and the sources that supplied hydrated fly ash to these districts.

Table 1. Fly Ash Sources for TxDOT Experimental Projects.

TxDOT District	Power Company	Plant/Station	Fly Ash Contractor
Amarillo	Southwestern Public Service Company (SPS)	Tolk Station, Muleshoe, TX	DePauw Fly Ash, Amarillo
Amarillo	Southwestern Public Service Company (SPS)	Harrington Station, Amarillo, TX	DePauw Fly Ash, Amarillo
Atlanta	SWEPCO	Welch Station, Cason, TX	Gifford-Hill Fly Ash, Dallas
Childress	Southwestern Public Service Company (SPS)	Harrington Station, Amarillo, TX	DePauw Fly Ash, Amarillo

Table 2 compares the chemical composition of fly ash obtained from the above three sources and Table 3 compares their physical properties.

Table 2. Average Chemical Composition of Fly Ash from Welch, Tolk and Harrington Power Plants as of 12/29/94 (TxDOT 1994).

	Welch Plant	Tolk Plant	Harrington Plant
Sum of SiO_2, Al_2O_3 and Fe_2O_3 (%)	50.47	55.12	56.24
CaO (%)	29.05	27.00	26.24
SO_3 (%)	3.84	1.87	1.87
MgO (%)	5.75	5.23	4.97
Moisture Content (%)	0.00	0.00	0.00
Loss on Ignition (%)	0.59	0.24	0.48
Available Alkalies (%)	1.34	1.41	1.53

Table 3. Average Physical Properties for Fly Ash from Welch, Tolk and
Harrington Power Plants as of 12/29/94 (TxDOT 1994).

	Welch Plant	Tolk Plant	Harrington Plant
Strength Activity Index (%)	104.29	105.58	109.38
Water Requirement (%)	94.92	94.52	94.77
Fineness (%)	12.88	16.90	22.16
Soundness (%)	0.013	0.024	0.025
Specific Gravity	2.7	2.65	2.66
Shrinkage (%)	0.004	0.009	-0.001

Review of the data presented in Tables 2 and 3 reveals that the unhydrated
fly ash material obtained from all 3 sources are very similar in terms of their
chemical composition and their physical properties.

LABORATORY CHARACTERIZATION OF HYDRATED FLY ASH

As mentioned previously, one of the two primary objectives of this study
involved the characterization of HFA in the laboratory. The primary goal of this
experimental program was to investigate whether HFA meets the TxDOT
requirements for flexible base material as shown in Table 4 for their best quality
(Grade I) flexible base material.

Table 4. TxDOT Specifications for Grade I Flexible Base Materials (TxDOT
1994).

Test / Property	TxDOT Specification Requirement
Texas Triaxial Test:	
Minimum Compressive Strength at 0 kPa Confining Pressure	310 kPa
Minimum Compressive Strength at 104 kPa Confining Pressure	1210 kPa
Gradation:	
Cumulative Percent Retained on	
45 mm Sieve	0
22 mm Sieve	10-35
10 mm Sieve	30-50
No. 4 Sieve	45-65
No. 40 Sieve	70-85
Wet Ball Mill Test:	
Wet Ball Mill Value	40
Max. Increase in Passing #40 Sieve after Test	20

This section describes the test methods used as well as data obtained from the tests conducted on hydrated fly ash from SPS Tolk Power Plant.

Unconfined Compressive Strength Test

Unconfined compression strength tests were conducted to find out how the compressive strength of HFA develops with age at different hydration (curing) water contents. Hydration water content is the water content used in the hydration process expressed as a percent by weight of dry powdered fly ash. Cylindrical specimens with a diameter of 7.6 cm (3.0 in.) and a height of 15.2 cm (6.0 in.) were prepared at 20, 40, 60, and 100 percent hydration water contents. Specimens were tested after 3, 7, 14 and 28 days of curing. Three replicates were made for each test condition.

Fly ash and distilled water were weighed in required proportions and mixed for 30 seconds using a rotary mixer at 450 rpm. The mixing time was selected to ensure thorough mixing of powdered fly ash and water. The mix was then placed in cylindrical molds and slightly vibrated to ensure proper placement without any voids (honeycombs) being formed in the specimen. Once all specimens were cast, polyethylene bags were used to cover the molds to avoid moisture loss and the specimens were kept in the laboratory at 70°F. Specimens were removed from molding about 15 minutes prior to testing to enable them to reach laboratory ambient conditions. Once demolded, it was observed that these specimens had a very good finish except for a couple of them that had some delamination damage close to the top surface during removal from molding. Test results from such specimens were discarded. It was also observed that specimens with 100 percent hydration moisture content were found to be too soft to test even after 28 days. Data from the unconfined compressive strength tests are shown in Figure 1.

Figure 1 indicates that the compressive strength decreases significantly with increasing hydration water content. The difference between the compressive strength of specimens with 20 and 40 percent hydration water contents is extremely large. It can also be seen that the strength gain reached an asymptotic level at a time nearing 4 weeks for all hydration water contents.

Wet Ball Mill Test

The Wet-Ball Mill Test was conducted according to TxDOT Standard Test Procedure Tex-116-E (TxDOT Manual of Test Procedures, 1995) on HFA produced in the laboratory and HFA produced at the Tolk power plant. The objective of this test was to observe the effect of hydration moisture content on degradation of HFA. The failed compressive strength test specimens with hydration moisture contents of 20 and 60 percent were further crushed to obtain aggregates used for this test. The gradation of the hydrated fly ash was adjusted to match that of the hydrated fly ash supplied by the Tolk Power Station. The results of these tests, with three replicates, are summarized in Table 5 below. Once again, the sensitivity of the properties of HFA to the hydration water content is quite evident.

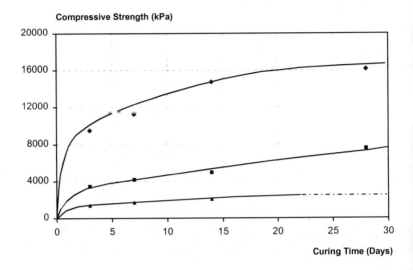

Figure 1. Unconfined Compressive Strength of Hydrated Fly Ash at Different Hydration Moisture Contents and Curing Times.

Table 5. Wet Ball Mill Test Data

HFA Source	Hydration Moisture Content (%)	Wet Ball Mill Value	Increase in Material Passing No. 40 Sieve (%)
Laboratory	20	23.00	14.00
	60	70.32	61.32
Tolk Station	-	39.43	36.55

Unit Weight Test

The objective of this test was to find out how the unit weight of hydrated fly ash varies with the hydration moisture content. For this test, an additional cylindrical specimen similar to those cast for the compressive strength test was used. These specimens were allowed to cure at 70°F and 50% relative humidity. Weight and geometric dimensions of specimens with different hydration water contents was measured and unit weight calculated. The results obtained are shown in Table 6.

Table 6. Specific Gravity of Hydrated Fly Ash at Different Hydration
Moisture Contents

Hydration Moisture Content (%)	Unit Weight (g/cm^3)
20	1.85
40	1.4
60	1.13
100	0.88

Compaction Test

The objective of this test was to determine the maximum dry density and the optimum moisture content, and the test was conducted on HFA from Tolk power station. Two gradations of HFA, well-graded and gap-graded, were tested. TxDOT Standard Test Procedure Tex-113-E described in the TxDOT Manual of Testing Procedures was used for this test. The results are shown in Table 7.

Table 7. Compaction Test Results

Gradation of Aggregate	Optimum Moisture Content (%)	Maximum Dry Density (g/cm^3)
Well-graded	24.0	1.34
Gap-graded	20.2	1.32

Electron Microscopy Studies

The electron microscopy studies were conducted on both fly ash and HFA. It was hoped that the microstructure of hydrated fly ash would provide clues that would help explain the variability in the performance of HFA as reported by different TxDOT districts.

Visual examination of the cylindrical HFA test specimens cast in the laboratory revealed that there were three zones distinguishable by color in the hydrated fly ash specimens. These are identified in Figure 2. Since this observation may give some clues as to the formation of different compounds under different hydration conditions, it was decided to conduct electron microscopy studies on these three regions. Fly ash and several samples of fly ash hydrated under different conditions were included in this study.

White Deposit on Surface

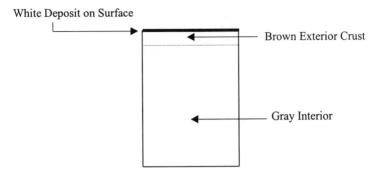

Figure 2. Three Distinct Hydration Zones in Hydrated Fly Ash Specimens.

The electron microscopy studies involved two types of analysis. They were: (a) Scanning and Transmission Electron Microscopy (STEM) Analysis, and (b) Scanning Electron Microscopy Analysis (SEM).

(a) Scanning and Transmission Electron Microscopy (STEM) Analysis:

Qualitative elemental analyses were carried out using the STEM to identify the elements found in fly ash and HFA.

(b) Scanning Electron Microscopy Analysis (SEM):

The development of the microstructure and the hydration products of fly ash were monitored for different hydration conditions.

Preparation of HFA Samples for Electron Microscopy

To prepare HFA samples for microscopy, fly ash was mixed with water for 30 seconds using a mechanical mixer. Test specimens were prepared from fly ash hydrated at water contents of 30, 40, 60, and 100 percent. HFA specimens were then cured under different conditions. Some specimens were kept sealed using polyethylene sheets and others were cured under laboratory conditions (70°F and 50 percent relative humidity). Then the specimens were tested at the ages of 10, 21, 28 and 42 days.

At the age of 10 days the samples hydrated at 60 and 100 percent moisture content showed white spots on the outer surface. The SEM micrograph of these white spots revealed the formation of rod-like ettringite crystals on the surface of hydrated fly ash. This is shown in Figure 3(a). The same sample at the age of 42 days was seen with a white fungus like growth on the outer surface and the SEM micrograph of this growth showed well-crystallized particles.

The samples hydrated at 20, 30 and 40 percent moisture content did not show such a phenomenon on the surface. Instead SEM micrographs of these specimens indicate the formation of compounds which are similar to monosulfoaluminate and calcium silicate hydrates formed in Portland cement

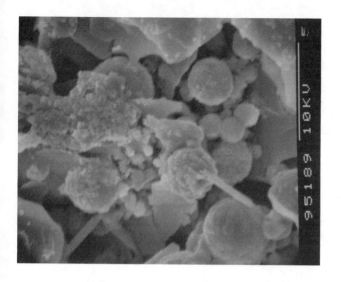

(b)

(a)

Figure 3.   Scanning Electron Micrographs of Hydrated Fly Ash; (a) SEM of the outer surface of hydrated fly ash at 100% moisture content, (b) SEM of the outer surface of hydrated fly ash at 40% moisture content

hydration as shown in Figure 3(b). The SEM micrographs of the inside surface of the specimens also indicate monosulfoaluminate and calcium silicate hydrates formation. A more complete discussion on all the micrographs corresponding to different hydration moisture contents and curing times can be found in Nash et al. (1996).

The important observations from the microscopical examination can be summarized as follows:

1. Hydration water content has a strong influence on the microstructure of HFA.
2. Curing conditions appear to have an influence on the hydration products.
3. Microstructure of HFA is different in the specimen interior and in regions exposed to the environment.
4. Particle size distribution of fly ash seems to have an influence in the hydration process. Larger particles take longer time to react with water or react partially or remain inactive. This can be observed from the micrograph of hydrated fly ash with 100 percent moisture content and at the age of 42 days.

PERFORMANCE OF HYDRATED FLY ASH IN ACTUAL PAVEMENT CONSTRUCTION PROJECTS

As mentioned previously, TxDOT constructed several experimental projects where HFA was used as the flexible base material. These experimental pavement sections were located in Amarillo, Atlanta and Childress districts. HFA for these experimental pavement sections were obtained from three different sources identified in Table 1. The specific construction projects completed in each TxDOT district and the experience from the use of hydrated fly ash is documented below.

Amarillo District

TxDOT constructed several experimental pavement sections in the Amarillo district. HFA for these projects came from the Harrington and Tolk power plants of the Southwestern Public Service (SPS) power company. The fly ash contractor for SPS is DePauw Fly Ash and DePauw of Amarillo. They maintain curing pits adjacent to the power plants and oversee the production of HFA.

In the Amarillo district, TxDOT experience with HFA has been positive. However, TxDOT engineers pointed out difficulties in meeting existing TxDOT flexible base specifications on gradation. When this material was placed and compacted, it crushed to smaller particles thus altering the gradation at the construction site. But once the HFA flexible base was placed, it hardened into a very stiff layer.

It was also observed that HFA required a significantly higher quantity of water for compaction. The optimum moisture contents for the fly ash from Harrington and Tolk power plants ranged from 20 to 30 percent. In spite of the high compaction water contents, there was no difficulty in achieving the required compaction densities.

Childress District

In the Childress district, TxDOT constructed two sections of highway detours on US-287 using hydrated fly ash as flexible base materials. These sections were in service for six to eight months and reportedly, their performance had been even better than with conventional flexible base materials.

Atlanta District

In the Atlanta district, TxDOT constructed five experimental pavement sections using HFA as flexible base. HFA for these projects came from the Welch power plant in Cason, operated by SWEPCO power company. Fly ash from SWEPCO is being marketed by Gifford-Hill Fly Ash of Dallas. Information from TxDOT indicated that the Welch power plant produced 585 Mg of fly ash per day and that Gifford-Hill Fly Ash has the capability to crush 1080 tons of fly ash per day. It was estimated that the available supply in the curing pit was approximately 335,000 cubic meters. Their curing pit occupied an area of 18 acres with fly ash stockpiled 4.6 meters deep.

Based on tests performed by TxDOT on this HFA, it was noted that the problem areas for using HFA are its high liquid limit (around 40) and the high optimum moisture content for compaction (25 to 30 percent). The advantages were identified as a low unit weight resulting in economical haulage cost, high strength, ease of achieving the required degree of field density and the availability of a perpetual supply of fly ash.

TxDOT engineers in Atlanta district identified two problems associated with pavements with seal coated surfaces containing HFA in base. One is the stripping away of the seal coat from the hydrated fly ash flexible base, particularly in the acceleration and deceleration zones. It was noted that there appear to be a lack of bond between the HFA base and the seal coat. It was also noted that a white crystalline material formed near areas of pavement where the seal coat was peeled off and the base layer was exposed to environment. These problems were not observed in pavements where hot mix asphalt concrete (HMAC) was used in the surface layer.

This investigation also revealed that the HFA used in the Atlanta District were produced under less controlled conditions when compared with those used in other districts. The authors believe that the problems identified in the Atlanta district may be associated with the use of higher hydration water contents. It is reasonable to conclude that the white crystalline material that appeared on the surface of the exposed base course is the same substance that was noticed on the outer surface of laboratory specimens that were hydrated higher water contents.

CONCLUSIONS

The findings from this study indicated that, from the strength standpoint, hydrated ash has great potential to be used as a pavement base material. However, the findings also show that unless the hydration conditions are carefully controlled, durability problems may arise.

Several properties of HFA including strength, resistance against degradation and microstructure were investigated in this study. In addition, data from TxDOT sources wer also evaluated and the following conclusions were drawn.

1. HFA is quite strong when compared to TxDOT specification triaxial classes. It easily meets the TxDOT Grade I flexible material requirements very easily.

2. Fly ash hydrated at lower water contents provide much higher strengths resulting in be resistance of the aggregate to degradation. Also, thorough mixing to achieve uniform distribution of water is very important.

3. HFA produced with higher water contents has lower unit weight and lower strength. These also showed a fungus-like growth on the surface when exposed to moisture in t field.

4. Sufficient curing time (a minimum of two weeks) is also important for good performa Therefore, care must be taken during the field curing process to allow sufficient time between addition of water and crushing of the hydrated product. If crushing is carried too early, the material may not meet the requirements for degradation and durability.

5. Hydrated fly ash undergoes further hydration after placement, thus forming a stiff and nearly homogeneous layer. Therefore, strict adherence to the gradation specification not be necessary.

ACKNOWLEDGMENTS

The work described in this paper was conducted as a part of a research study sponsored b Texas Department of Transportation (TxDOT).

REFERENCES

Bloomquist, D., G. Diamond, M. Oden, B. Ruth and M Tia, "Engineering and Environm Aspects of Recycled Materials for Highway Construction," *Final Report FHWA-RL*

088, Western Research Institute, Laramie, Wyoming, 1993.Manual of Testing Procedures, Texas Department of Transportation, Austin, Texas, 1995.

Grant, M., Personal Communication, District Laboratory Supervisor, Texas Dept. of Transportation, Amarillo District, 1995.

Nash, P.T., P.W. Jayawickrama, S.P. Senadheera, J. Borrelli and A.S.M. Rana, "Guidelines for Using Hydrated Fly Ash as a flexible Base," Final Report No. 0-1365-1F, Texas Department of Transportation, Austin, Texas, 1996.

Pollard, T.J.S, M.D. Montgomery, and J.C. Sollar, "Organic Compounds in the Cement-Based Stabilization/Solidification of Hazardous Mixed Waste – Mechanistic and Process Consideration, Research Report, Imperial College Center for Toxic Waste Management, Imperial College of Science, Technology and Medicine, London, U.K., 1992.

anders, J.F., T.C. Keener and J. Wang, "Heated Fly Ash/Hydrated Lime Slurries for SO_2 removal in Spray Dryer Absorbers," *Indian Eng. Chem. Res.*, Vol. 34, No.1, New Delhi, India, 1995.

Texas Dept. of Transportation, *Manual of Testing Procedures*, Materials and Tests Division, Austin, Texas, 1995.

Texas Dept. of Transportation, *Standard Specifications for Construction of Highways, Streets and Bridges*, Austin, Texas, 1994.

Texas Dept. of Transportation, *Material Test Data Sheets*, Materials and Tests Division, Austin, Texas, 1994.

Repeated Loading of Stabilized Recycled Aggregate Base Course

Khaled Sobhan[1] and Raymond J. Krizek[2]

ABSTRACT

An experimental investigation was conducted to study the fatigue behavior, resilient properties, and progressive accumulation of damage due to repeated flexural loads on a fiber-reinforced pavement base course material composed of cement-stabilized recycled concrete aggregate and fly ash. The primary objectives of this endeavor were (a) to determine the resistance to fatigue failure in terms of traditional S-N curves, and compare the behavior with other typical stabilized materials, (b) to determine the variation of the cumulative plastic strain with the number of loading cycles and the degradation of the dynamic elastic modulus, (c) to evaluate the extent of accumulated damage in terms of a fatigue damage index, and (d) to demonstrate the application of laboratory-derived material properties in a mechanistic design method. All specimens contained (by weight) 4% cement, 4% fly ash, and 92% recycled aggregate; the fiber-reinforced specimens contained an additional 4% (by weight) hooked-end steel fibers. Results show that the unreinforced material has a fatigue strength comparable to virtually all typical stabilized highway materials. The degradation of the dynamic elastic modulus due to repeated loading was found to be less than 25% of the initial modulus. The resilient modulus in flexure for this material was found to be comparable to values reported for traditional soil-cements materials. Miner's Rule of linear summation of damage is applicable to the unreinforced material, but not to the fiber-reinforced material. A modest amount of reinforcing fibers was quite effective in increasing the fatigue resistance and retarding the rate of damage accumulation in this lean cementitious composite containing primarily waste materials. Finally, the laboratory derived fatigue properties are incorporated into an elastic layer mechanistic method to illustrate a typical base course design.

INTRODUCTION

It is generally recognized that the utilization of recycled materials in civil engineering construction, such as a highway pavement, is a promising concept from both

[1]Assistant Professor of Civil Engineering, Bucknell University, Lewisburg, PA 17837. Phone: (717) 524-1491; email: sobhan@bucknell.edu
[2]Stanley F. Pepper Professor of Civil Engineering, Northwestern University, Evanston, IL 60208

environmental and economic standpoints. However, a careful characterization of the candidate waste materials is necessary to assess their suitability as a component in the pavement structure. The current study addresses the potential use of cement-stabilized waste concrete aggregate as a base course for highway pavements. Since a stabilized pavement layer is subjected to repeated flexural (tensile) stresses due to moving wheel loads which will ultimately cause fatigue failure in the pavement, an experimental investigation was undertaken to study the flexural fatigue strength and the gradual accumulation of permanent damage in the material due to repeated loads. To date, a number of studies have been reported on the laboratory fatigue characterization and/or design aspects of soil-cement or aggregate-cement base course materials (Larsen and Nussbaum, 1967; Mitchell and Shen, 1967; Carpenter et al., 1992; Thompson, 1994); a few studies were also conducted on fiber reinforced soil-cement (Craig et al., 1987; Crockford et al., 1993; Maher and Ho, 1993). Cavey et al. (1995) conducted coordinated laboratory and field experiments on the utilization of waste fibers to reinforce a base course composite consisting of stabilized recycled aggregate, and   Sobhan and Krizek (1996) reprted results from unconfined compression, split tensile, and static flexural tests on stabilized recycled aggregate reinforced with commercially available fibers. Recently, Sobhan and Krizek (1998) presented experimental findings about the fatigue damage and degradation properties of this composite. However, no information was found in the literature on (a) the fatigue strength and endurance limits of this composite as compared to other traditional pavement materials, or (b) the mechanistic design of such pavement layers utilizing dynamic material properties obtained from laboratory studies.

OBJECTIVES
The specific objectives of this study on a stabilized base course material composed largely of recycled aggregate and fly ash are to:
(a)   evaluate the flexural fatigue behavior of a cement-stabilized recycled aggregate base course material and compare its performance with traditional stabilized pavement materials,
(b)   investigate the extent of fatigue damage in terms of accumulated plastic strains and degradation of the elastic modulus with the number of loading cycles,
(c)   calculate a fatigue damage index in terms of cumulative dissipated energy, and
(d)   demonstrate the application of laboratory-derived material properties in a mechanistic design approach

MATERIALS
The recycled concrete aggregate was obtained from the R. I. Busse Corporation in Elk Grove, Illinois. The grain size distribution of this material is such that 100%, 93%, 72%, 63%, 38%, and 3% of the material passes the 1-inch, 3/4-inch, 1/2-inch, 3/8-inch, No. 4, and No. 200 sieves, respectively. Ordinary Type I Portland cement and fly ash complying with ASTM C618 specifications were used as stabilizing materials. The fly ash had a specific gravity of approximately 2.7 and was obtained from the American Fly Ash Company in Naperville, Illinois. Commercially available hooked-end (HE) steel fibers (trade name Bekeart Dramix), with a diameter of 0.8 mm and a length of 60 mm, were used to reinforce the stabilized base course composite.

SPECIMEN PREPARATION AND CURING

A total of seventeen beams, each having dimensions of 152.4 mm x 152.4 mm x 686 mm (6 in. x 6 in. x 27 in.), were prepared; five were tested in static flexure and twelve under repeated flexural loads. Two of the statically tested beams and six beams subjected to fatigue tests were fiber reinforced. Each beam consisted of 92% recycled aggregate, 4% cement, and 4% fly ash by weight; the fiber-reinforced specimens contained hooked-end steel fibers at a dosage of 4% by total dry weight of the mix. All test specimens were prepared at a target dry density of 1826 kg/m^3 (114 pcf) by using a static compaction technique (Sobhan and Krizek, 1996) to simulate the slow roller compaction procedure used in the field. Specimens were cured for 28 days in a 100% relative humidity room.

TEST CONFIGURATION

Beams were tested in bending under a third-point loading configuration over a span length of 584 mm (23 inches) using an MTS closed-loop testing system. All fatigue tests were conducted under load control using a nonreversed haversine load pulse with a constant amplitude (which was different for each specimen) at a frequency of 2 Hz (120 cycles/minute). The dynamic mid-span deflection was measured with a 7.6 mm (0.3-inch) linear variable differential transformer. A high-speed external data acquisition system was used to continuously record the cyclic load deformation data during each experiment for all loading cycles until failure.

EXPERIMENTAL RESULTS

The magnitude of the repeated flexural stress applied to each beam was selected on the basis of the results obtained from a static flexural test program which was reported by Sobhan and Krizek (1996). A conventional way of presenting the results of a series of fatigue tests is to plot the so-called S-N curve which is the relationship between the stress ratio (ratio of applied flexural stress to static flexural strength) and the number of cycles to failure. The procedure for estimating the flexural strengths of individual beams for the purpose of calculating the applied stress ratios are described in the reference (Sobhan, 1997). The magnitude of the repeated flexural stress, static flexural strength of each individual beam, stress ratio, and the number of loading cycles to failure are presented in Table 1. In this table, the six unreinforced beams are designated as FU1 through FU6 and the six fiber-reinforced beams are designated as FR7 through FR12. Beam FU2 failed during the first cycle at a flexural strength of 1000 kPa (145 psi); therefore, a repeated load test could not be performed on this beam. Beams FU5 and FR10 did not fail after 2 million cycles of loading. The average flexural strength of the unreinforced specimens is 1489 kPa (216 psi) and that of the fiber-reinforced specimens is 1612 kPa (234 psi), an increase of only 8%, implying that the fibers (at the dosage used) do not improve the strength meaningfully.

THE S-N CURVES AND COMPARISON STUDIES

The relationship between the stress ratio and the number of cycles to failure (S-N curves) for the unreinforced and fiber-reinforced specimens are shown in Figure 1 by the two solid lines. One of the objectives of the fatigue experiments was to compare the performance of the recycled aggregate base course with that of similar stabilized base

TABLE 1. Results of Repeated Load Test Program

Beam	Flexural Strength (MPa)	Stress Ratio (SR)	Cycles to Failure (N_f)	Resilient Modulus at 5% Fatigue Life (MPa)	Dynamic Elastic Modulus at 99% Fatigue Life (MPa)	Degradation of Elastic Modulus (%)
FU1	1.492	0.62	432000	3570	2770	23
FU3	1.381	0.71	10197	5060	3000	41
FU4	1.424	0.69	5636	5300	4630	13
FU5	1.477	0.63	2000000	10410	NA[a]	NA[a]
FU6	1.670	0.62	31496	3710	3400	9
FR7	1.625	0.87	337	4270	3250	24
FR8	1.626	0.83	118	2820	2380	16
FR9	1.610	0.83	17246	8960	2980	67
FR10	1.641	0.79	2000000	9210	NA[a]	NA[a]
FR11	1.681	0.80	2218	4910	2500	49
FR12	1.479	0.87	12688	8550	5010	42

[a]Deformations were not measured up to failure

course materials. Five such materials were chosen for this purpose: (a) high strength stabilized base (HSSB) materials, (b) concrete, (c) lime- fly ash-aggregate, (d) soil-cement, and (e) lean concrete. Typical S-N relationships for these materials have been retrieved from the literature (Carpenter et al., 1992) and superimposed on Figure 1 with dashed curves. These comparisons show that the performance of unreinforced recycled aggregate is quite similar to that of soil-cement, and both materials perform better than HSSB, concrete, and lime-fly ash-aggregate beyond the 100-cycle region. The performance of fiber-reinforced recycled aggregate is much superior to all other materials (except lean concrete) in the low-to-high cycle range. The curve for the fiber-reinforced recycled aggregate is flatter than all other curves, indicating greater resistance to fatigue damage and degradation compared to unreinforced materials. Since all recycled aggregate beam specimens in this study had approximately similar strengths, as shown in Table 1, the superior performance of the fiber-reinforced specimens under repeated loads can only be attributed to the addition of fibers. A nonlinear regression analysis on the data for recycled aggregate specimens gives the following relationships between the stress ratio, SR, and the fatigue life, N:

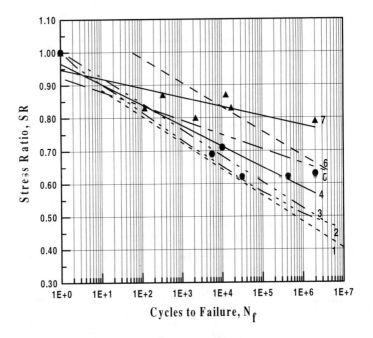

Figure 1. **Stress Ratio versus Number of Cycles to Failure for Various Stabilized Pavement Materials: (1) High Strength Stabilized Base Materials, (2) Concrete, (3) Lime-Fly Ash, (4) Unreinforced Recycled Aggregate, (5) Soil-Cement, (6) Lean Concrete, and (7) Fiber-Reinforced Recycled Aggregate**

(a) For unreinforced specimens (R^2=0.91):

$$SR = -0.028 \ln(N) + 0.965 \tag{1}$$

(b) For fiber reinforced specimens (R^2=0.64):

$$SR = -0.012 \ln(N) + 0.944 \tag{2}$$

The 2-million-cycle fatigue endurance limits for the unreinforced and fiber-reinforced materials (calculated from the above equations) are, respectively, 56% and 77% of the static modulus of rupture. These results suggest that a relatively inexpensive base course containing recycled aggregate with low quantities of cement is very promising in terms of fatigue strength; furthermore, the inclusion of a modest amount of reinforcing fibers to this lean composite brings about a significant improvement in its resistance to fatigue failure.

ANALYSIS OF CYCLIC LOAD-DEFLECTION DATA
To study the gradual accumulation of damage due to repeated loads, complete load-deformation data were acquired for all loading cycles to failure, and these data are analyzed in the following sections.

Accumulation of Permanent Strain
In a typical load-deformation cycle the total vertical deformation, Δ_T, is the sum of an elastic or recoverable deformation, Δ_R, and a permanent deformation, Δ_P:

$$\Delta_T = \Delta_R + \Delta_P \tag{3}$$

The permanent strain at the end of each cycle can be computed from the observed permanent deformations (Sobhan and Krizek, 1998); the variation of the cumulative permanent strain with the number of load cycles is an indication of the progressive damage sustained by the material. The accumulated permanent strain for various specimens is plotted against the cycles ratio (number of cycles normalized by the total number of cycles to failure) in Figure 2. As a comparison, Figure 2(h) shows the curve for a 152.4 mm x 152.4 mm x 533.4 mm (6 in. x 6 in. x 21 in.) concrete beam which failed at 836,990 cycles when loaded to a stress ratio of 0.6 (Tawfiq, 1994). This figure shows that all of the curves have approximately similar shapes and the rate of plastic strain accumulation exhibits three distinct stages:

(a) **Stage I** : In this initial stage there is a rapid growth of damage, as depicted by the fast accumulation of permanent strain; for most specimens shown in this figure, Stage I is approximately up to 5% of the fatigue life.
(b) **Stage II** : Stage II is an intermediate stage in which the rate of permanent strain accumulation approximately stabilizes, indicating predominantly elastic deformation; for the specimens shown, Stage II consists of the region between 5% and approximately 90% of the fatigue life.
(c) **Stage III** : This is the final phase, during which the permanent strain accumulates at an accelerating rate towards failure; Stage III consists of the region between 90% and 100% of the fatigue life.

Dynamic Elastic Modulus
A mechanistic pavement design method requires the input of an elastic modulus (termed resilient modulus) determined from repeated load tests. The general equation for calculating the resilient modulus, RM, from flexural tests is (Vinson, 1990):

$$RM = \frac{K P}{I \Delta_R} \tag{4}$$

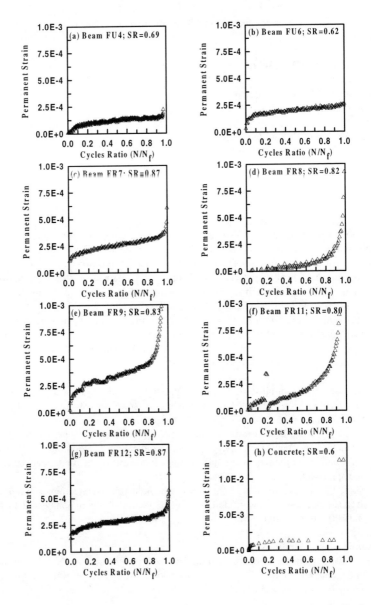

Figure 2. Variation of Accumulated Permanent Strains with Cycles Ratio

where K is a constant depending on test configuration and specimen geometry (in this study, $K = a(3L^2 - 4a^2)/48$, where L is the length of the span, and $a = L/3$), P is the magnitude of repeated load, I is the moment of inertia of the cross-section, and Δ_R is the dynamic recoverable deflection at the midspan. In this study, Δ_R was measured at 5% of the fatigue life, which corresponds to the beginning of the predominantly elastic region (Stage II), as described in the previous section. The resilient moduli calculated in this fashion are given in Table 1. Assuming that the fibers (at the dosage used) do not influence the strength or the modulus (as shown by Sobhan, 1997), the specimens can be divided into two groups: (a) low modulus specimens, having resilient moduli ranging approximately between 2.75 GPa (400,000 psi) to 5.5 GPa (800,000 psi), and (b) high modulus specimens, having resilient moduli values ranging approximately between 8.27 GPa (1.2 million psi) to 10.3 GPa (1.5 million psi). Mitchell and Shen (1967) determined the resilient modulus of soil-cement in flexure and presented a relationship between the flexural strength and the resilient modulus in a log-log plot similar to that shown in Figure 3. Superimposed on their results are the values determined from the recycled aggregate beam specimens in this study. The flexural strengths reported in Table 1 were used to plot the points for the recycled aggregate specimens. A best fit regression curve through all of the points has the following form ($R^2 = 0.95$):

$$\ln(RM) = 1.758 \ S_f + 5.84 \tag{5}$$

where S_f is the flexural strength. Since the recycled aggregate specimens had almost similar strengths, they form a cluster about the best fit curve. It is found that the resilient modulus values for stabilized recycled aggregate are comparable to the values of traditional soil-cements.

In general, Equation 4 can be used to compute a dynamic elastic modulus, E_d, for each cycle of loading by replacing Δ_R with the observed elastic deformation at the end of that cycle. The degradation of the elastic modulus is also an indication of damage accumulation due to fatigue. Variations of the dynamic elastic modulus with the cycles ratio have been presented by Sobhan and Krizek (1998). The computed elastic modulus at 99% of the fatigue life, and the total degradation of the modulus (expressed as a percentage of the corresponding resilient modulus computed at 5% fatigue life) for each specimen is presented in Table 1. It is observed that, for most of the specimens, the total degradation of the elastic modulus varied between 9% and 24% at or near failure. Although it was found that the elastic modulus degrades slowly with loading cycles (Sobhan and Krizek, 1998), it is not unreasonable in most cases to assume a constant resilient modulus for purposes of conducting a linear elastic analysis.

Energy Approach to Fatigue Damage Characterization
It was observed during the repeated load tests that the loading and unloading paths in each cycle were typically different, which produced a hysteresis loop. The area within the hysteresis loop denotes the plastic strain energy which is dissipated or "lost". The

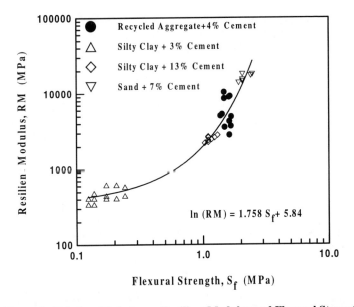

Figure 3. Relationship between Resilient Modulus and Flexural Strength

total energy dissipation capacity (which is the sum of the areas of all the hysteresis loops up to failure) at a given stress or strain level is a material property, and the specimen will fail when this capacity is exceeded. Detailed studies of the size and shapes of hysteresis loops as found in this experimental program, are available in the reference (Sobhan, 1997). As reported by Sobhan and Krizek (1998), there is a strong correlation between the magnitude of the total dissipated energy and the fatigue life for this material. In view of this correlation, a damage index, D, can be defined as the ratio of the cumulative energy dissipated for a given number of cycles, E, to the total energy dissipation capacity, E_{tot}, of the material at a given stress level:

$$D = \frac{E}{E_{tot}} \tag{6}$$

This definition of damage was proposed and used by Grzybowski and Meyer (1993) for concrete tested in compression. Figures 4(a) and 4(b) show the variation of the damage index, D, with the cycles ratio for unreinforced and fiber-reinforced specimens, respectively. The straight line drawn at 45^0 is called Miner's Rule (1945) for cumulative damage and is often used in pavement engineering; it is expressed as $D_i = \Sigma (n_i/N_i)$, where N_i is the number of cycles to failure at stress level i and n_i is the actual number of cycles applied at that stress level. The ratio n_i/N_i represents the damage done due to stress level i; it is assumed that the damage accumulates linearly in the material, which

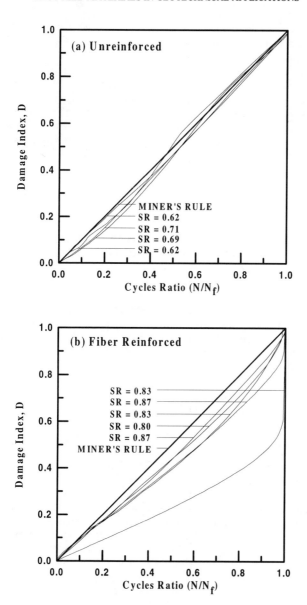

Figure 4. Variation of Damage Index with Cycles Ratio

fails when the linear sum of damage is unity. It is observed from Figure 4 that the damage accumulation in unreinforced specimens at stress ratios ranging from 0.6 to 0.7 shows a slight nonlinearity up to about 40% of the fatigue life, after which it closely follows Miner's Rule. Therefore Miner's Rule is applicable to unreinforced material in this range of stress ratios. However, all fiber-reinforced specimens show a quite pronounced nonlinear accumulation of damage at stress ratios ranging from 0.8 to almost 0.9, and therefore Miner's Rule cannot be used in this range of the stress ratio to predict the remaining life of a pavement made from fiber-reinforced materials. It was also found that there is a noticeably lower accumulation of damage in fiber-reinforced specimens, although they were subjected to significantly higher stress ratios compared to the unreinforced specimens. Therefore, a modest amount of reinforcing fibers is effective in retarding the rate of damage accumulation due to repeated loads. Figure 4 can be used directly to assess the degree of damage sustained by the stabilized pavement layers in the above ranges of stress ratio.

MECHANISTIC-EMPIRICAL DESIGN CONSIDERATIONS

A mechanistic design approach involves the input of material properties and traffic load information into a theoretical structural model which can be used to calculate pavement response in terms of stresses, strains, and deflections. These responses are then correlated to pavement performance, as measured by the type and severity of distress (e.g. rutting and cracking), by means of suitable transfer functions; these transfer functions are often derived empirically from laboratory data and are generally calibrated by observing the performance of in-service pavements. The laboratory derived fatigue and resilient properties, as described in this study, are utilized in this section to illustrate these mechanistic design principles in the case of a pavement containing a stabilized recycled aggregate base course and a relatively thin asphalt surfacing. In this type of pavement, the subgrade stresses and the asphalt concrete tensile strains are low, and the pavement derives most of its structural capacity from flexural (tensile) strength of the stabilized base (Thompson, 1994).

As a design example, several hypothetical pavement sections are analyzed with a mechanistic computer program named ELSYM5 (Kopperman et al, 1986), which is based on linear elastic layered theory. Each pavement section consists of (a) a 102-mm (4-inch) asphalt surface with a resilient modulus, RM, of 2.75 GPa (400,000 psi) and a Poisson's ratio of 0.35, (b) a subgrade with a resilient modulus of 20.7 MPa (3,000 psi) and a Poisson's ratio of 0.4, and (c) a stabilized recycled aggregate base course with a resilient modulus of 3.45 GPa (500,000 psi) and a Poisson's ratio of 0.25. The base course thickness was varied from 127 mm to 152 mm (5 inches to 10 inches) for the purpose of this analysis. The selected base course moduli fall within the ranges obtained in the current experimental program. The pavement response, which in this case is the maximum tensile stress at the bottom of the stabilized base layer, is calculated due to a 40 kN (9000-lb) wheel load applied on a circular area with a tire contact pressure of 690 kPa (100 psi). These results are shown in Table 2. The computed tensile stresses are divided by the average flexural strengths of 1489 kPa (216 psi) for unreinforced materials and 1612 kPa (234 psi) for fiber-reinforced materials to calculate the applied stress

Table 2. Design Calculations for Unreinforced and Fiber-Reinforced Materials

Thickness of Base Course (mm)	Tensile Stress at Bottom of Base (kPa)	Unreinforced Base Course		Fiber Reinforced Base Course	
		Stress Ratio (SR)	Cycles to Failure (N)	Stress Ratio (SR)	Cycles to Failure (N)
127	1482	≈ 1.0	1	0.92	8
152	1193	0.8	362	0.74	24×10^6
178	979	0.66	53790	0.61	NF
203	814	0.55	2.7×10^6	0.5	NF
229	683	0.46	NF	0.42	NF
254	586	0.4	NF	0.36	NF

Note: NF = Stress ratio was too low to cause fatigue failure

ratios. The performance of a pavement can then be predicted in terms of the number of cycles to failure by using the laboratory derived fatigue transfer functions presented in Equations 1 and 2; these results are also shown in Table 2. The effect of the variable base course thickness on the predicted pavement performance is presented in the design chart shown in Figure 5, which illustrates that, for any desired number of cycles to failure, the use of fiber-reinforced material will result in a significant reduction in the base course thickness compared to unreinforced material. A series of similar curves can be prepared for other values of the base course resilient modulus and for different thicknesses and properties of the asphalt and subgrade layers.

SUMMARY AND CONCLUSIONS
The study reported herein was undertaken to evaluate the behavior of a cement-stabilized recycled aggregate base course material subjected to repeated flexural stresses which simulate the effect of idealized traffic loadings on highway pavements. The results of the experimental program show that the proposed base course material manifests good resistance to fatigue failure and that a modest amount of suitable fibers improves the mechanical characteristics which are critical to pavement loading conditions. Following are the significant conclusions of this investigation:

1. The fatigue behavior (as depicted by traditional S-N curves) of an unreinforced recycled aggregate base course material containing only 4% cement and 4% fly ash is comparable to virtually all commonly used stabilized pavement materials. The addition of 4% by weight (1% by volume) of hooked-end fibers significantly enhances this fatigue resistance. The 2-million- cycle endurance limits for unreinforced and fiber-reinforced material are, respectively, 56% and 77% of the static modulus of rupture.

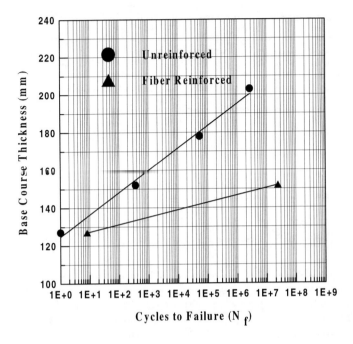

Figure 5. Variation of Base Course Thickness with Cycles to Failure

2. During repeated flexural loading, permanent strain accumulates in the material in three stages: (a) in Stage I (0 to 5% of fatigue life) there is a rapid growth of damage with a fast accumulation of strain; (b) during Stage II (5% to 90% of fatigue life) the accumulation of strain is slow and relatively stable; and (c) Stage III (90% of fatigue life to failure) is a period of accelerated strain accumulation resulting in failure.

3. The resilient modulus in flexure varies between a low value of 2.75 GPa (400,000 psi) and a high value of 10.3 GPa (1,500,000 psi); these values are comparable to those of traditional soil-cements in flexure. For most specimens, the dynamic elastic modulus degrades slowly under repeated loading, with the maximum deterioration not exceeding 25% at or near failure.

4. Miner's Rule of linear summation of damage may be applicable to unreinforced specimens with stress ratios ranging from about 0.6 to 0.7, but a nonlinear accumulation of damage precludes its applicability to stabilized pavement materials with fiber reinforcement for stress ratios above about 0.8.

5. The use of 4% (by weight) fibers was effective in retarding the rate of fatigue damage accumulation in the proposed base course material, which was not only lean in

cement (4% by weight) but was composed of 96% (by weight) waste materials, such as recycled aggregate and fly ash.

6. The use of fiber reinforcement may allow significant reduction in the base course thickness according to a mechanistic design of a typical flexible pavement.

REFERENCES

Carpenter, S. H., Crovetti, M. R., Smith, K. L., Rmeili, E., and Wilson, T. (1992). "Soil and Base Stabilization and Associated Drainage Considerations: Volume I, Pavement Design and Construction Considerations," Report No. FHWA-SA-93-004, *Federal Highway Administration.*

Cavey J. K., Krizek, R. J., Sobhan, K., and Baker, W. H. (1995). "Waste Fibers in Cement-Stabilized Recycled Aggregate Base Course Material," *Transportation Research Record, No. 1486,* Transportation Research Board, Washington D. C., pp. 97-106.

Craig, R., Schuring, J., Costello, W., and Soong, L. (1987). "Fiber Reinforced Soil Cement." *American Concrete Institute*, SP-105, ed. S. P. Shah, and G. B. Batson.

Crockford, W. W., Grogan, W. P., and Chill, D. S. (1993) "Strength and Life of Stabilized Layers Containing Fibrillated Polypropylene." Paper No. 930888, 72nd Annual Meeting, Transportation Research Board, Washington, D. C.

Grzybowski, M., and Meyer, C. (1993). "Damage Accumulation in Concrete with and without Fiber Reinforcement," *ACI Materials Journal*, Vol. 90, No. 6, pp. 594-604.

Kopperman, S., Tiller, G., and Tseng, M. (1986).   "ELSYM 5: Interactive Microcomputer Version," *User's Manual*: IBM-PC and Compatible Version, Report No. FHWA-TS-87-206, Federal Highway Administration.

Larsen, T. J., and Nussbaum, P. J. (1967). "Fatigue of Soil-Cement." *Journal of the Portland Cement Association*, Vol. 9, No. 2, pp. 37-59.

Maher, M. H., and Ho, Y. C. (1993). "Behavior of Fiber-Reinforced Cemented Sand Under Static and Cyclic Loads." *Geotechnical Testing Journal*, American Society for Testing and Materials, Vol. 16.

Miner, M. A. (1945). "Cumulative Damage in Fatigue," *Transactions of the American Society of Mechanical Engineers*, Volume 67, pp. A159-A164.

Mitchell, J. K., and Shen, C-K. (1967). "Soil-Cement Properties Determined by Repeated Loading in Relation to Bases for Flexible Pavements," Second International Conference on the Structural Design of Asphalt Pavements, University of Michigan, August 7-11, pp. 348-364.

Sobhan, K. and Krizek, R. J. (1998). "Resilient Properties and Fatigue Damage in a Stabilized Recycled Aggregate Base Course Material." Paper No. TRB 980157. Presented at the 77th Annual Meeting of the *Transportation Research Board* in Washington, D. C., January 11-15, 1998 (in press).

Sobhan, K. (1997). "Stabilized Fiber-Reinforced Pavement Base Course with Recycled Aggregate," *Ph.D Dissertation*, Department of Civil Engineering, Northwestern University.

Sobhan, K. and Krizek, R. J. (1996). "Fiber Reinforced Recycled Crushed Concrete as a Stabilized Base Course for Highway Pavements," *First International Conference on Composites in Infrastructure*, University of Arizona, Jan.15-17, pp. 996-1011.

Tawfiq, K. (1994). "Fatigue Fracture in Concrete," *Final Report*, No. FL-DOT-RMC-0-0623-4068, Florida Department of Transportation.

Thompson, M. R. (1994). "High-Strength Stabilized Base Thickness Design Procedure," *Transportation Research Record No. 1440*, Transportation Research Board, Washington, D. C., pp. 1-7.

Vinson, T. S. (1990). "Fundamentals of Resilient Modulus Testing," *Proceedings of the Workshop on Resilient Modulus Testing*, Oregon State University, Publication No. FHWA-TS-90-031, Federal Highway Administration, Session I, pp. 27-59.

Compaction Characteristics of Contaminated Soils-Reuse as a Road Base Material

Jay N. Meegoda[1], Bin Chen[2], Samiddha D. Gunasekera[3] and Philip Pederson[4]

Abstract

Contaminated soils generated from leaking underground storage tanks (USTs) sites are classified as solid waste or non-hazardous waste. Since petroleum contaminated soils are solid waste they cannot be used as clean fill material. A sub-base of a major freeway in New Jersey was built with petroleum contaminated soils. The petroleum soil was appropriately contained to prevent leaching of petroleum products. In order to understand the compaction behavior of petroleum contaminated soils, an extensive laboratory study was conducted. The following conclusions were drawn from the test results. When the soils are contaminated with non-polar organic liquids, due to the lubricating action, there is an improvement in compaction characteristics. This is not continued beyond the oil content value corresponding to the situation where all the soil particles are coated with oil. When the soils are contaminated with polar organic liquids, besides the above lubricating action, the soil structure tends to be dispersed. The dispersed soil structure produced low maximum dry density. However, if the pore liquid viscosity was high the soil structure effect is masked to improve the compaction characteristics. A 2.5% glycerol content improved the compaction characteristics of all the clayey soils tested.

Introduction

During the 1950s and 1960s, the construction of many gasoline stations, chemical manufacturing and processing facilities led to the installation of millions of USTs. Several million USTs in the United States contain petroleum products. Petroleum contaminated soil (PCS) is generated from these leaking Underground Storage Tanks (USTs), and their piping systems. Tens of thousands of these USTs,

[1]Associate Professor and [2]Graduate Student, respectively, Dept. of Civil and Env. Eng., New Jersey Institute of Technology, Newark, NJ, Meegoda@Megahertz.NJIT.edu
[3]Geotechnical Engineer and [4]Senior Chemist, respectively, BEM Systems Inc., Florham Park, NJ

including their piping systems, are currently leaking (Fairweather, V., 1990). The United States Environmental Protection Agency (USEPA) estimated in 1988 that there were more than 400,000 leaking USTs with petroleum hydrocarbons (USEPA, 1988). Many more are expected to leak in the future. The free petroleum products released from USTs will eventually move to contaminate the groundwater. Since groundwater is a source of drinking water, federal legislation seeks to safeguard our nation's groundwater resources. Congress responded to the problem of leaking USTs by adding Subtitle I to the Resource Conservation and Recovery Act (RCRA) of 1984. The current federal and state statutes require such leaking USTs to be removed to prevent further contamination. Most states vigorously encourage the removal of all tanks after 25 years of service. It is estimated that the removal of a leaking tank on average generates approximately 50 to 80 cubic yards of contaminated soil. Soils from petroleum refineries and crude oil wells are also contaminated with petroleum products. Many sites identified through regulatory programs such as RCRA have soils for disposal that are not classified as hazardous waste, but as solid waste. Since petroleum contaminated soils are solid waste they cannot be used as clean fill material. Quantities of such soils are projected to increase substantially over the next few years. In addition, the availability of solid waste disposal facilities is limited. Petroleum contaminated soils consist of mixtures of natural sands, silts and clays with petroleum products.

By 1984, the State of New Jersey had generated 500,000 tons of PCS for disposal by removing leaking USTs (Britton, 1984). The disposal rate for such soil varies from $35.00 to $100.00 per ton depending on the quantity and location. There are several treatment options, for example, thermal treatment and biological degradation. Thermal treatment costs vary from $40.00 to $100.00 per ton, depending on the location and quantity. Biological treatment process is slightly less expensive but it takes time to process the PCSs. There were several attempts to reuse PCSs and return them to the economic mainstream as products. Ezeldin et al., 1992 proposed the use of PCSs as sand replacement for the production of structural concrete and used them in non-residential construction, that is, in the construction of highways, parking lots, sidewalks, etc. Meegoda et al., 1992 proposed the use of PCSs as aggregate replacement for the production of hot mix asphalt concrete. PCSs were also used as road base material through compaction to specifications and subsequently, covering them with nearly impermeable asphalt concrete or regular concrete layer to prevent leaching of petroleum products to the groundwater. Any of the soil reuses mentioned above will reduce the demand for solid waste disposal facilities by turning PCSs into usable products. The following section of this paper documents a field demonstration of the use of PCS as road base material.

Field Demonstration of Reuse of PCSs as Sub Base Material

The New Jersey Department of Transportation (NJDOT) recently completed the construction of the extension of Route I-287 in Northern New Jersey. It stretched from the Pequannock River in Bloomingdale, Passaic County northward to the Wanaque River in Passaic County, and eastward to the Ramapo River in Bergen County. Wetlands located within the right-of-way of this highway were destroyed and required to be replaced. A 10 acre industrial site, occupied by a concrete pipe manufacturer, was chosen as a replacement wetland. This site contained two concrete pipe manufacturing buildings, a truck maintenance facility and a drying kiln. Besides the above structures, there were two aboveground storage tanks and four underground storage tanks (USTs) used to store gasoline, kerosene and diesel oil. There was a drum storage area used to store waste oil and hydraulic oil. Site investigations showed that there was approximately 18,000 cubic yards contaminated soils. Soil contamination was caused by the use and on-site storage of fuels, and hydraulic oil required for operations of the concrete pipe manufacturer. The contaminated soils were excavated, a small percentage of highly contaminated soils (chemical concentrations greater than 30,000 ppm) was moved off-site, and the remainder was stockpiled in the immediate vicinity of the wetland site.

In the state of New Jersey, soil to be recycled or reused must be classified as non-hazardous according to the NJ administrative code 7:28-8. To characterize the soil, waste classification proposed by the New Jersey Department of Environmental Protection (NJDEP) was implemented. A total of 72 five part soil samples was taken from the contaminated soil stockpile for analysis. The predominant contaminants found in the soil samples were alkanes and polycyclic aromatic hydrocarbons consistent with petroleum sources. Results of the chemical analysis of the soil did not indicate the presence of any metals or organic chemicals with concentrations exceeding the minimum allowable concentration levels for hazardous waste classification. Soil analysis revealed that the total petroleum hydrocarbons (TPHCs) concentration ranged from 11.5 to 5,280ppm, with an average concentration of 1800 ppm. Samples found with TPHC concentrations greater than 1000 ppm (56 of 72 samples) were analyzed for priority pollutants. No sample with concentrations exceeding the NJDEP soil action level of 10 ppm for individual chemicals was found in the base/neutral analysis of the samples (BEM Systems 1990). Based on the test results of the soil samples, the NJDEP issued a ID-27 classification (industrial waste) for the stockpiled soil.

The decision to reuse the PCSs was based on a cost analysis of the operation. Initial cost estimates for soil disposal revealed that the incineration cost $8.1 to 13.5M, $2.2 to 2.7M for landfilling, and $0.15 to 0.2M for reuse as a road base material. Therefore, reuse as a road base material was found to be the most cost effective alternative. The reuse cost was essentially due to the processing of the soils. Transportation costs were minimal as the soil was already located near the site. Based on potential savings, the reuse option was implemented. Figure 1 shows a typical roadway section with contaminated soil used as construction material for the road base. During construction, contaminated soils were separated from the embankment edge by an 8-foot berm of uncontaminated zone embankment material. It was compacted in layers 8 inches to 12 inches deep until the required 2-foot depth was achieved. The contaminated soil was covered as soon as possible with an 8 inch compacted road base layer made of soil aggregates forming part of the roadway pavement. The NJDOT reused all of the contaminated soils excavated from the wetland mitigation site as roadway embankments fill material for Route I-287.

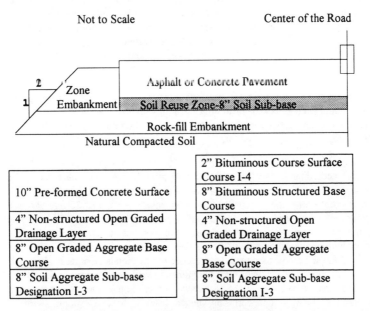

Type A Pavement Main Lane	Type B Pavement Shoulders
	2" Bituminous Course Surface Course I-4
10" Pre-formed Concrete Surface	8" Bituminous Structured Base Course
4" Non-structured Open Graded Drainage Layer	4" Non-structured Open Graded Drainage Layer
8" Open Graded Aggregate Base Course	8" Open Graded Aggregate Base Course
8" Soil Aggregate Sub-base Designation I-3	8" Soil Aggregate Sub-base Designation I-3

Type A Pavement Main Lane   Type B Pavement Shoulders

Figure 1. Typical Roadway Section with Contaminated Soils.

A site investigation comprising Geotechnical soil borings was carried out from 1982 to 1984 along Route I-287 from the Pequannock River to the Wanaque River. The results of this site investigation generally indicated the presence of surface soils of varying thickness consisting of sands with coarse to fine gravel and traces of silt overlying bedrock composed of hard, fractured gneiss. Also, the boring log results of the Geotechnical investigations carried out at Wetlands Mitigation Site No. 2, and a gradation performed on the soil stockpile revealed the following: about 12% of soil passed sieve #200 and soil had an in-situ moisture content of 8.5%. These data suggested that the fine content of the stockpile material was low. Consequently, relatively low levels of dust emission were considered likely to result from screening operations.

There was no treatment of the contaminated soils before its reuse, other than by screening of oversize material to remove boulders and stones of size greater than 4 inches. The boulders and stones removed during the screening operation were treated as non-contaminated material, subject to   approval by the NJDEP, and were either used on site within the roadway alignment as rock-fill embankment, or moved off-site to a secured landfill. Fugitive dust was controlled during screening operations by wetting the soil with water using a fine mist spray as required by NJDEP, both before screening and during screening operations. A nuisance dust level of 10 mg/m^3 of air was maintained as recommended in Threshold Limit Values and Biological Exposure Indices (American Conference of Governmental Industrial Hygienists, 1988-1989).

The case study discussed above was a comprehensive field demonstration. However, there were several unanswered geotechnical engineering questions such as the effects of the contaminants on compaction characteristics, and the governing mechanisms for compaction characteristics of contaminated soils. The following laboratory study was designed to address these issues.

Laboratory Investigation of Compaction Characteristics of Contaminated Soils.

Any attempt to study the compaction characteristics of contaminated soils, needs the complete understanding of the mechanism behind mechanical compaction. Compaction results could be explained through consideration of the effects of moisture content and density on shear strength; consolidation characteristics; and permeability of compacted soils. Existing theories of compaction such as those referenced below, have fulfilled these aims with only limited success.

Considerable research effort was spent on the compaction of soil, and many theories have been proposed.   There are a few well-accepted theories of compaction such as Proctor's Theory (Proctor, 1933), Hogentogler's Viscous Water Theory (Hogentogler, 1936), Lambe's Theory on Physico-chemical Interaction (Lambe, 1960) and Olson's Effective Stress Theory (Olson, 1963). These theories include various pore liquid parameters that affect the compacted dry density of soil.

From the above theories, it can be inferred that the following properties of the organic liquids should influence the dry density-liquid content relationship:
1. Viscosity: Two early theories referred to the lubrication provided by the liquid. During compaction, high viscous liquids provide good lubrication for the soil particles. This was used to overcome the inter-particle forces and slide against each other to form a denser soil structure. This lubricating effect was predominant at low liquid contents.
2. Surface tension: Surface tension of the liquid exerts a force on the soil particles, therefore resisting relative movement during compaction. However, this effect is not very significant and may be negligible around the optimum water contents and on the wet side of the optimum.
3. Dielectric Constant: The increase in the dielectric constant and the physico-chemical interaction of the pore liquid resulted in a decrease in the shear strength of the soil (Lambe, 1958). The reduced shear strength facilitated the

easy sliding of soil particles on one another to a denser arrangement, hence a higher dry density.

4. Liquid Density: There is no information on the influence of liquid density on compaction characteristics. However, it may be reasonable to assume that with the increase in liquid density there is a corresponding increase in liquid content (water content).

Therefore, the following experimental investigation was proposed and conducted for the detailed evaluation of influence of above pore liquid properties on compaction characteristics of contaminated soils.

Experimental Investigation

Two artificial soils and one natural soil were used in this experimental study. These soils were selected such that they represent a clay of high plasticity, a clay of low plasticity, and a silty clay. The natural soil used was a silty clay from the Brunswick Formation, North New Jersey (referred to as soil #2). Two artificial soils were commercially available kaolin (soil #1), and a mixture of commercially available Bentonite (15% by weight) and kaolin (85% by weight) (soil # 3). A grain size analysis and Atterberg limit tests were performed on each soil to identify and classify them. The liquid limit, plastic limit, and hydrometer analysis tests were performed on the three soils. The results obtained for the above tests are shown in Table 1. The soils #1, #2, and #3 were classified as a low plastic clay, a silty clay, and a high plastic clay respectively.

Table 1. Measured Soil Properties of the Clays.

Soil	Liquid Limit (%)	Plastic Limit (%)	Specific Surface Area (m^2/g)	Percent Finer than 2 μm.
Kaolin	48	36	55	84
Mixed Soil	73	35	128	88
New Brunswick Soil	37	34	98	18

The three pure chemicals used in this investigation were Glycerol, 1- Propanol, and Acetone. All three liquids were water soluble and represent organic liquids of medium to high dielectric constants (20-40). In addition, the water soluble chemicals are non-toxic. Their proportions in the solutions were selected such that they cover a full range of possible mixtures. The Propanol percentage was limited to 50% due to a limitation in electrical conductivity; glycerol to 44% due to a limitation in viscosity; and acetone to 37.5% due to a limitation in evaporation and safety. The three chemicals were mixed with de-ionized water to obtain the desired solutions and sufficient amounts of NaCl (salt) were added to bring the solution conductivity to $5.0*10^{-4}$ mhos/cm. This was done to eliminate influences of the ion type and amount on the soil structure. The above salt concentration levels were also similar to those in the water used for the field compaction of soils. The physical properties of the liquids and the mixtures of liquids considered in this investigation are listed in Table 2. For the initial qualitative study, commercially available SAE 10W-30 motor oil was used. Properties of this oil are similar to the properties of heating oil, a common contaminant. Physical properties of this oil are not listed.

Table 2. Measured Physical properties of the Organic Chemicals.

Liquid	Viscosity (centi poise)	Surface Tension (dynes/cm)	Dielectric Constant	Density (kg/m^3)
Acetone	0.337(15)	23.7(20)	20.7(25)	789.9(20)
1-Propanol	2.52 (15)	23.78(20)	20.1 (25)	803.5(20)
Glycerol	2330 (15)	63.4 (20)	42.5 (25)	1261.3(20)
Water	1.339(15)	73.05(20)	80.37(25)	1000(4)

Note: Values shown within brackets are temperature values (in ^oC)

The compaction test for each sample was performed according to ASTM testing procedure D-1557. The only deviation from the established ASTM test procedure was the use of chemical mixtures instead of water in the compaction test. The term "liquid content" is used in this text, instead of "water content," for technical accuracy (Meegoda and Ratnaweera, 1987). For the soil sample mixed with water, the standard compaction test was performed.

For the initial qualitative tests, soil #1 was used. This test series was qualitative as there was no information on motor oil. Also, since motor oil is immiscible in water, it is difficult to perform a quantitative analysis. In this test series the following were used: soil #1 and soil #1 uniformly mixed with 3%, 6% and 12% motor oil respectively. When the soil was mixed with oil, it was simply treated as contaminated soil. Water was added incrementally to perform the compaction test regardless of the contamination level of the soil. Four separate compaction tests were performed on soil #1, soil #1 with 3% oil, soil #1 with 6% oil and soil #1 with 12% oil. The test results are plotted in Figure 2. Please note that in Figure 2 instead of "water content" the word "liquid content" is used. The liquid content was determined by heating the soil to 200°C instead of 105°C to remove all the oil from the soil. If soils were not heated beyond 200°C, there was no change in Geotechnical properties of soils on heating (Wang et al., 1990). Therefore, it was decided to heat soil to 200°C to determine the liquid content. It is noteworthy that the difference in liquid contents, when heated to 105°C and 200°C, is essentially the oil content in the soil.

With increase in oil contamination, there was an initial increase (3% and 6%) and a decrease (12%) in maximum dry density. This increase was assumed to be due to the lubricating action of oil. To further investigate this phenomenon, a quantitative investigation was performed. In this investigation, three water-soluble organic liquids were used.

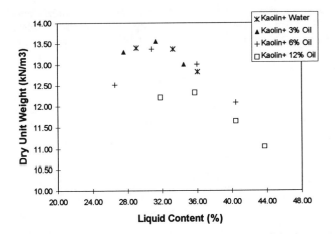

Figure 2. Compaction Test Results of Soil #1 and Soil #1 Contaminated with Oil.

For the quantitative test series, kaolin was first used with different solutions of glycerol to perform the compaction test. Different sets of compaction curves were obtained for each pore liquid as shown in Figure 3 and Table 3. The compaction tests were repeated, taking care to strictly adhere to ASTM test procedure (ASTM D-1557), and they superposed the same test data as shown in the Figure 3. Similar test results were obtained for mixed soil with Propanol (see Figure 4 and Table 3) and New Brunswick soil with acetone (see Figure 5 and Table 3).

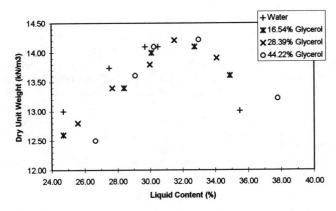

Figure 3. Compaction Test Results for Soil #1 with Water and Glycerol Solutions.

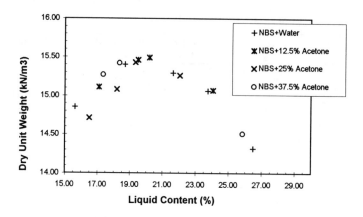

Figure 4. Compaction Test Results for Soil #2 with Water and Acetone Solutions.

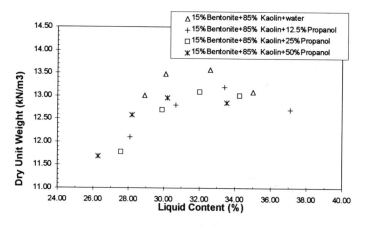

Figure 5. Compaction Test Results for Soil #3 with Water and Propanol Solutions.

The experimental results showed the following for kaolin with glycerol; there was a significant increase in dry density or the compaction characteristics. It was assumed that the very high viscosity of glycerol improved the lubrication properties of the liquid. Though the dielectric constant of glycerol is less than that of water the effect of the lubrication seems to dominate. When the proportion of glycerol in the liquid increased, the maximum dry density of the soil also increased. The above test results indicated the following: for the same compaction effort, the maximum dry density of a fine soil can be increased at a slightly lower water content by the addition of glycerol.

Table 3. Compaction Test Data for the Three Soils.

Soil Type	Pore Liquid Type	Maximum Dry Unit Weight [kN/m^3]	Optimum Liquid Content [%]
Soil #1	Water	14.12	30.4
Soil #1	16.54% Glyc.	14.16	31.4
Soil #1	28.39% Glyc.	14.22	31.6
Soil #1	44.22% Glyc.	14.33	32.0
Soil #2	Water	15.42	19.7
Soil #2	12.5% Ace.	15.46	20.8
Soil #2	25.0% Ace.	15.49	21.5
Soil #2	37.5% Ace.	15.50	21.7
Soil #3	Water	13.70	31.5
Soil #3	12.5% Pro.	13.30	33.0
Soil #3	25.0% Pro.	13.20	33.0
Soil #3	50.0% Pro.	13.10	31.0

Table 4. Compaction Test Results for Soil #1 with Glycerol Water mixes.

% Glycerol	0.0	0.3	1.0	2.5	5.0
Maximum Dry Density (kN/m^3)	14.12	14.15	14.20	14.22	14.22
Optimum Liquid Content (%)	31.0	30.0	29.5	29.5	29.5

Table 5. Compaction Test Results for Soil #2 with Glycerol Water mixes.

% Glycerol	0.0	1.0	2.5
Maximum Dry Density (kN/m^3)	16.87	16.38	16.87
Optimum Liquid Content (%)	22.0	20.0	18.5

Table 6. Compaction Test Results for Soil #3 with Glycerol Water mixes.

% Glycerol	0.0	1.0	2.5
Maximum Dry Density (kN/m^3)	13.73	13.73	13.83
Optimum Liquid Content (%)	31.5	30.5	31.0

Additional compaction tests were performed on other two soils with different amounts of glycerol in water to further investigate the possibility of improving the compaction characteristics of the soils. The test results are shown in Tables 4, 5, and 6.

Discussion of Test Results

Figure 2 clearly indicates the following: the maximum dry density of kaolin increased from 1365 kg/m^3 to 1385 kg/m^3 (or a 1.5% increase), and the optimum liquid content decreased from 32% to 28.2% with 3% oil contamination. When the oil content increased from 3% to 6% the maximum dry density did not increase but there was an increase in optimum liquid content to 30.9%. However, when the oil content increased from 6% to 12% the maximum dry density reduced to 1260 kg/m^3 and the optimum liquid content increased to 34%. From these test results up to a maximum of 4.5% oil content, there appears to be an increase in compaction characteristics with the increase in oil content. Meegoda and Ratnaweera (1995) showed that soil tends to show granular characteristics with increase in organic liquid content in the soil. Johnson and Sallberg (1960) showed that with the increase in granular nature of soil, there was an increase in maximum dry density or the compaction characteristics of the soil. However, if this theory is used, then with increase in oil content, compaction characteristics should show a constant improvement. That was not the case in this research. Hence, it was assumed that with the increase in oil content, there was an increase in the lubricating effects that produced the higher maximum dry densities. These lubricating effects disappeared (at an oil content of 4.5%) when soil particles were fully coated with oil. It was apparent that some oil (at oil contents higher than 4.5%) occupied the void spaces hindering the compaction process. The study with soluble organic liquids validated the influence of pore liquid viscosity on soil compaction.

The maximum dry unit weight of the kaolin varied from 14.1 kN/m^3 (90.0 pcf) for the water compacted sample to 14.3 kN/m^3 (91.3 pcf) for the sample compacted with 44.2% glycerol solution. The optimum liquid content increased from 30.6% to 32.3% when the amount of glycerol increased. All the measured liquid content values were corrected for the different liquid densities in order to compare all the test results with zero air voids line. Since organic solutions were used in place of water, for the determination of liquid content the term "water content" was modified to "equivalent water content." This was done to take into account the difference in specific gravities of these solutions. The new term may be explained using the following two assumptions. A liquid with a density greater than water is used in place of it to determine the liquid content of the soil. The void ratio corresponding to the liquid content value with the liquid is the same as that with water. However, due to the higher density of the liquid, the liquid content value will be higher than that with water. If the equivalent water content is used as defined below, then the contribution of liquid density on liquid content may be accounted. Then, the compaction curves can be drawn with the zero air void line for the soil.

$$\text{Equivalent w/c= liquid content } \frac{(c+1)}{(c+G_f)} \text{ ------------ (1)}$$

where c is the ratio of volume of water to volume of chemical in the pore liquid, and G_f is the specific gravity of the pure chemical. If a pure chemical was used, then the value of c is zero.

The compaction curves were plotted again with the equivalent water content and are shown in Figure 6 and summarized in Table 7. All the curves merged on the wet side of the compaction and were parallel to the zero air voids line. However, the maximum dry unit weight gradually increased, and the optimum equivalent water content decreased as the degree of contamination increased. This observation may be viewed as obtaining better compaction for the same compactive effort with the increasing degrees of contamination. The results for soil #2 and soil #3 (see Figures 7 and 8) showed that the maximum dry density reduced and optimum moisture content increased when water solutions were used instead of water.

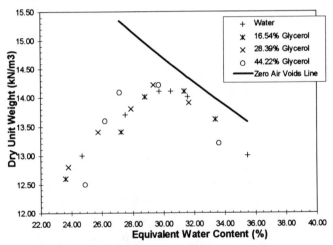

Figure 6. Compaction Data for Soil #1 with Glycerol Solutions with ZAV Line.

Different soil structures are formed when polar organic liquids are mixed with soils. This is due to the physico-chemical interactions of soil-liquid-system. With lower dielectric constant of the pore liquid there is a weaker physico-chemical interaction that causes soils to exhibit dispersed soil structure. With a dispersed soil structure it is difficult to produce a dense matrix with compactive action. Hence it produces lower maximum dry densities. On the contrary, the viscosity of the pore liquid plays a major role in improving the compaction characteristics. If the pore liquid viscosity is higher than that of water the pore liquid, then the liquid tends to lubricate the soil causing an improvement in compaction characteristics. For soil #1 with glycerol viscous effects of glycerol dominated to improve the

the pore liquid viscosity is higher than that of water the pore liquid, then the liquid tends to lubricate the soil causing an improvement in compaction characteristics. For soil #1 with glycerol viscous effects of glycerol dominated to improve the compaction characteristics. For soil #2 and #3 the reduction can be attributed to the comparatively low dielectric constants and viscosities of Acetone and Propanol when compared with Glycerol.

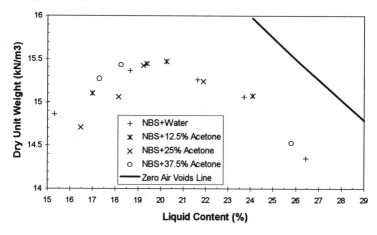

Figure 7. Compaction Data for Soil #2 with Acetone Solutions with ZAV Line.

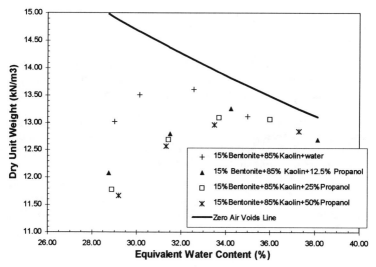

Figure 8. Compaction Data for Soil #3 with Propanol Solutions with ZAV Line.

Tables 4, 5, and 6 show that by the addition of glycerol improved the compaction characteristics of all three soils. The above tables also show that the optimum glycerol content would be around 2.5% to show an improvement in compactive effort considering the additional cost of glycerol.

Table 7.   Soil Compaction Data for Three Soils with Various Pore Liquids.

Soil Type	Pore Liquid Type	Maximum Dry Unit Weight [kN/m³]	Optimum Liquid Content [%]	Equivalent Water Content [%]
Soil #1	Water	14.12	30.4	30.4
Soil #1	16.54% Glyc.	14.16	31.4	30.1
Soil #1	28.39% Glyc.	14.22	31.6	29.4
Soil #1	44.22% Glyc.	14.33	32.0	28.7
Soil #2	Water	15.42	19.7	19.7
Soil #2	12.5% Ace.	15.46	20.8	21.4
Soil #2	25.0% Ace.	15.49	21.5	22.7
Soil #2	37.5% Ace.	15.50	21.7	23.5
Soil #3	Water	13.70	31.5	31.5
Soil #3	12.5% Pro.	13.30	33.0	33.8
Soil #3	25.0% Pro.	13.20	33.0	34.7
Soil #3	50.0% Pro.	13.10	31.0	34.4

Summary and Conclusions

There is high emphasis placed on transportation agencies to reuse the contaminated soils encountered during highway expansion projects or new highway projects. A case study is reported, where the contaminated soils were used as construction material for road base.

A detailed laboratory investigation was performed to understand the mechanics behind the compaction of contaminated soils. It was observed that when clay soils are contaminated with non-polar petroleum products, they behave as a granular (silty) soils with lower optimum moisture contents and higher dry densities. This behavior of plastic soil was investigated in this study to improve compaction characteristic of clay soils. This new road base material will be subsequently covered by nearly impermeable asphalt concrete or regular concrete layer to prevent leaching of petroleum products to the groundwater.

When the soils are contaminated with non-polar organic liquids, due to the lubricating action there is an improvement in compaction characteristics. This increasing trend is not continued when oil content increases beyond the value where all the soil particles are coated with oil.

When the soils are contaminated with polar organic liquids, besides above lubricating action, the soil structure plays a major role. With the increase in contamination, the soil structure tends to be dispersed. The dispersed soil structure produces low maximum dry density. However, if the pore liquid viscosity is high, then the soil structure effect is masked to improve the compaction characteristics. A 2.5% glycerol content improved the compaction characteristics of all the clayey soils.

References

Britton, C. L. (1984), "New Jersey Ground-Water Contamination Index," Sept. 1974-April 1984, New Jersey Geological Survey, Open File Report #84-1.

Casagrande, A., (1948), "Classification and Identification of Soils", Transactions, ASCE, Vol. 113, pp. 901-930.

Ezeldin, A. S., Vaccari, D. A., Bradford, L., and Mueller, R. T., (1992) "Stabilization and Solidification of Hydrocarbon Contaminated Soils in Concrete", Journal of Soil Contamination, Vol. 1 #1, pp. 61-79.

Fairweather, V., (1990) "U.S. Trackless Leaking Tanks" Civil Engineering, Vol. #60, pp. 46.

Hogentogler, C. A., (1936) "Essentials of Soil Compaction", Proceedings , Highway Research Board, National Research Council, Washington DC, pp. 309-316.

Johnson, D. E., and Sallberg, J. R., (1960), Factors that Influence Field Compaction of Soils", Bulletin 272, Highway Research Board, 206pp.

Lambe, T. W., (1960), Structure of Compacted Clay", Transactions, ASCE, Vol. 125, pp. 682-705.

BEM Systems, Inc, (1990), "Soil Reuse Plan for Contaminated Soil Stockpile", A report presented to the New Jersey Department of Environmental Protection and Energy, June 1990.

Meegoda, N. J., and Ratnaweera, P. (1987) "A New Method to Characterize Contaminated Soils," Superfund '87, Washington, D. C., November 16-18, pp. 385-399.

Meegoda, N. J., and Ratnaweera, P., (1995), "Treatment of Oil Contaminated Soils for Identification and Classification", ASTM Geotechnical Testing Journal, Vol. 18, #1, March 1995, pp. 41-49,

Meegoda, N. J., Huang, D. R., DeBose, B., and Mueller, R. T. (1992) "Use of Petroleum Contaminated Soils in Asphalt Concrete", Hydrocarbon Contaminated Soils Volume II, Chapter 31, pp. 529-548. (P. T. Kostecki, E. J. Calabrese & M. Bonazountas, Editor), Lewis Publisher.

Olson, R. E., (1963) " Effective Stress Theory of Soil Compaction" ASCE Journal of SMFE, Vol. 89, #2, pp. 29-45.

Pederson, P., and Nardolillo, M., "Alternative Uses for Petroleum Contaminated Soils as Roadway Construction Materials", Superfund 1991, Washington DC, December 1991, pp. 719-722.

Proctor, R. R., (1933) "Fundamental Principles of Soil Compaction" ENR, Vol. 111, #9, pp. 245-248.

Wang, M. C., Benway, J. M., and Arayssi, A. M., (1990), "The effect of heating on Engineering properties of clays", Physico-Chemical aspects of soil and related materials, K. B. Hoddinott and R. O. Lamb (eds), ASTM STP 1095, pp. 139-158.

DATABASE ON BENEFICIAL REUSE OF FOUNDRY BY-PRODUCTS

Tarek Abichou[1], Craig H. Benson[2], and Tuncer B. Edil[3]
Members ASCE

Abstract: This paper describes a database on beneficial reuse of foundry by-products that was assembled at the University of Wisconsin-Madison. The database was designed to identify (1) materials and markets where methods and specifications exist such that beneficial reuse can begin or be expanded immediately and (2) areas in need of further research. Ninety projects in 14 states and two locations in Canada were evaluated. A broad variety of applications were identified including flowable fills, geotechnical fills, Portland cement concrete and related products, asphalt concrete, Portland cement manufacturing, landfill barrier layers, and agronomic amendments. A technical review was performed of several projects in each reuse area, which consisted of summarizing research activities and actual projects. Whenever possible, specifications and methods were identified and recommendations summarized.

INTRODUCTION

Even though many states have developed beneficial reuse regulations for industrial by-products, large quantities of foundry by-products are still being landfilled throughout the U.S. However, beneficial reuse of foundry byproducts is increasing in a variety of areas including the transportation, construction, and environmental industries. These industries can use large quantities of foundry by-products for earth fills, hydraulic control barriers, and pavement system aggregates. These applications are usually suitable for large foundries, where significant quantities of byproducts are available. Other reuse applications exist for smaller foundries, such as production of bricks, blocks, Portland cement, agricultural soil amendments, and other innovative applications usually related to local markets.

Information and experience gained through many beneficial reuse projects are not readily available to different sectors of industry. To help remedy this problem, a database was compiled describing projects throughout the US that have employed

---

[1] PhD Candidate, Dept. of Civil and Environ. Engr., University of Wisconsin-Madison, Madison, WI 53706, Ph. (608)262-6281, abichou@students.wisc.edu

[2] Assoc. Prof., Dept. of Civil and Environ. Engr., University of Wisconsin-Madison, Madison, WI 53706, Ph. (608)262-4272, chbenson@facstaff.wisc.edu

[3] Prof., Dept. of Civil and Environ. Engr., University of Wisconsin-Madison, Madison, WI 53706, edil@engr.wisc.edu, Ph. (608)262-3225

foundry by-products. The database includes projects that have not been reported in the literature, as well as those previously published. The database was designed to identify (1) materials and markets where methods and specifications exist such that beneficial reuse can begin or be expanded immediately and (2) areas in need of further research. This paper represents an overview of the database and the key findings in relation to the reuse of foundry by-products.

DATABASE DEVELOPMENT

The study consisted of the following tasks: (1) contact information sources, (2) compilation of application documents, (3) detailed technical review, and (4) identification of topics in need of research.

Contact Information Sources

The first task consisted of conducting an extensive literature review using the electronic library system at the University of Wisconsin-Madison, the Internet, the American Foundrymen's Society (AFS) library in Des Plaines, Illinois, and the Penn State foundry by-products list. Once the literature review was complete, emphasis was placed on obtaining unpublished information about projects where foundry by-products have been beneficially used. Representatives from departments of transportation (DOTs) and environmental regulatory agencies in the key foundry states (Wisconsin, Michigan, Illinois, Iowa, Indiana, Minnesota, Pennsylvania, Ohio, California, Texas, Louisiana) were interviewed to obtain information about projects involving foundry by-products, current DOT policies toward beneficial reuse of foundry by-products, and the regulatory status of the beneficial reuse of foundry by-products. Foundry organizations in all of the key foundry states were contacted and asked if they had listings of projects where foundry by-products had been beneficially used.

Results of this effort showed that most states have blanket policies permitting certain uses of foundry by-products. Listings of particular projects usually were not available either from state agencies or foundries, but some projects were verbally identified. Thus, much of the information was obtained through referrals obtained from the interviewees. However, several sources were not willing to share information about projects because of concerns about future liability. Nevertheless, 90 projects were identified in 14 states and two locations in Canada.

Compilation of Application Documents

The information collected in the first task was used to build a technical resources library that can be used in future studies regarding the beneficial reuse of foundry by-products. The information was also used to build a database of projects where foundry by-products have been used.

Microsoft Access® was used to build the database. The database permits immediate electronic access to all projects. Each project has a Project ID Form, which provides information about the project. Information on each ID form consists of (1) the name of project, (2) the type of application, (3) the type of material used, (4) the location, (5) a contact person, (6) a brief description of the project, and (7) a list of available references. Users can use Microsoft Access to search for a certain project, projects of similar application, and projects at the same location. The database can be updated as new information becomes available.

The types of applications described in the database are:

- *Structural/Base/Sub-baseFill*: projects where foundry by-products have been beneficially used as structural fill, embankment material, granular backfill, roadway sub-base, and roadway base material.
- *Flowable Fill*: projects where foundry by-products have been used in the production of controlled low strength materials (i.e., flowable fills).
- *Concrete and Related Products*: projects where foundry by-products have been used in the production of concrete bricks, pre-cast concrete such as blocks, and construction of concrete pavements.
- *Asphalt:* projects where foundry by-products have been used in asphaltic concrete.
- *Soil Amendments:* projects where foundry by-products have been used in agricultural applications.
- *Portland Cement:* projects where foundry by-products have been used in the production of Portland Cement.
- *Landfill Liners and Covers:* projects where foundry by-products have been used in landfill liner or cover construction.
- *Other Applications:* The database contains other applications such as pipe bedding, clay bricks, landfill drainage layers, etc.

Technical Review

A technical review of all significant projects involving beneficial reuse of foundry by-products was performed as the third task. All technical reviews began with a summary of research activities, followed by a review of actual projects. Since transportation and construction applications are likely to generate significant impacts on the beneficial reuse of foundry by-products, projects involving these applications were carefully reviewed. Whenever possible and appropriate, specifications and methods have been identified and recommendations have been summarized.

Identify Topics in Need of Research

Based on the compilation of application documents and the technical reviews, topics in need of research and development were identified and prioritized. Issues requiring investigation were identified for each application, and recommendations were made regarding research that should be conducted to fill the information gaps to further advance the beneficial reuse of foundry by-products.

The major findings of this effort are summarized in the next sections. The summaries are presented by reuse application and consist of an example of a typical technical review of a beneficial reuse project followed by specification and recommendations. A compilation of all technical reviews can be found in Abichou et al. (1998a).

FLOWABLE FILLS

Flowable fills are self-leveling liquid-like materials that cure to the consistency of a stiff clay. Benefits of flowable fills include limited required labor, accelerated construction, ready placement at inaccessible locations, and the ability to be manually re-excavated. Applications for flowable fills include utility trenches, building excavations, underground storage tanks, abandoned sewers and utility lines, slab jacking, and filling underground mine shafts (Smith 1996, Naik and Shiw 1997). Flowable fill is typically a mixture of sand, fly ash or cement, and water. Since sand is the major component of flowable fills, replacing the natural sand with foundry sand is an attractive beneficial reuse application. Javed (1994) concludes that when fly ash and foundry sand are supplied free of charge, flowable fills containing foundry sand can be produced at lower cost than with traditional materials.

Example Project

Flowable fill containing foundry sand was placed in trenches at the west abutments of the overpass at the intersections of County Highways D and G in Sheboygan County, Wisconsin in Fall 1996. Trenches at the east abutments were constructed using base course material (38 mm and 19 mm road gravel) as a regular backfill, and compacted in lifts using plate compactors and backhoe buckets. Comparison of the two fill materials is being based on pavement distress and pavement profiles and will be continued for three years.

Ready-mix trucks were able to effectively fill the entire excavation. No spreading was required since the material flowed well. Slump tests showed 250 mm of spread. The flowable fill mix used on this project consisted of 1560 kg of foundry sand, 240 kg Class C fly ash, 30 kg cement, and 422 L of water per cubic meter of flowable fill. Crowns to match the existing bridge decks were constructed by placing additional material on top of the flowable fill. Afterwards, base course was placed. Elevations at the centerline and the edge line at both approaches of the two structures were then obtained and used as the base line profiles for the structures. The first WisDOT interim report on the evaluation of this project concluded that, as of August 8, 1997, there is no difference between the performance of the flowable fill and the granular backfill.

Specifications

Ohio DOT specifications refer to flowable fills as low strength mortar backfills. They recommend the following mix design:

Table 1. Ohio DOT Specifications for Types I and II Flowable Fill.

Component	Type I	Type II
Cement (kg/m^3)	30	59
Fly Ash (kg/m^3)	148	0
Sand (kg/m^3)	1726	1436
Water (kg/m^3)	297	125 to 178

The Ohio specifications specifically mention the possibility of using foundry sands and require the development of alternative mixes to meet the strength and flowability criteria. These specifications also provide detailed information about mixing and placement of flowable fills.

The State of Wisconsin is developing specifications for flowable fills. The specifications require that the mix have a flow of 225 mm per ASTM PS-28-95 and a 28-day compressive strength of 280 to 550 kPa. The specification allows the use of natural sand, natural gravel, produced sand, foundry sand, fly ash, Portland cement, and other broken or fragmented mineral materials.

The Iowa Department of Transportation (IDOT) is a leader in the use of flowable fills. IDOT specifications require that the sand be fine with a maximum of 10% fines, the cement content not exceed 50 kg/m^3, and the total amount of cementitious material not exceed 250 kg/m^3. IDOT also developed the following mix for flowable fills containing foundry sand:

- 1300 kg foundry sand
- 250 kg fly ash
- 50 kg cement
- 1080 L water

These quantities are per cubic yard of flowable fill.

Recommendations

Use of foundry sand in flowable fills is a promising application. Foundry sand can be used as a partial or full replacement of natural sand. Use of foundry sand may also enhance the economy of flowable fills relative to conventional backfills. Design mixes should be developed to satisfy local and state strength and flow requirements. The mixes in Table 2 can be used as a starting point.

Table 2. Recommended Starting Mixes

Component	Mix 1	Mix 2 (IDOT)
Cement (kg/m^3)	30	59
Fly Ash (kg/m^3)	240	295
Foundry Sand (kg/m^3)	1534	1534

Water and cement quantities should be adjusted to meet flowability and strength criteria. Some specifications may also require a particular gradation of fine aggregates, which may require processing of foundry sand. Currently, there are no recommendations for the characteristics of the foundry sand; e.g., bentonite content; however there is a consensus that foundry sands behave better than concrete sands in flowable fills. Foundries should make data sheets describing their sand (including particle size distribution and bentonite content) available to contractors, ready-mix producers, and designers.

HIGHWAY EMBANKMENT AND SUB-GRADE

Foundry sands have been used as structural fill for many years. This practice ceased in the U.S. when new environmental laws were introduced. However, Ham et al. (1981) conclude that foundry wastes from ferrous foundries are generally non-hazardous. This finding has allowed several states to issue exemptions or blanket policies for the beneficial reuse of foundry wastes in structural fill applications. Most states, however, place restrictions on locations of such applications and require some type of encapsulation of the foundry sand.

Example Project

Mast (1997) describes an embankment demonstration project sponsored by the Indiana Cast Metal Association (INCMA), Auburn Foundry, and the Indiana Department of Transportation (InDOT). This project built on earlier work performed by Javed and Lovell (1994), and consisted of building a 300-m section of Route 206 near Butler, Indiana. Two control sections were constructed. One control section was built with clay typically used in the area; the other control section was built using clean sand. A section using foundry sand from the Auburn Foundry was built next to the two control sections.

Mast (1997) presents field data from the embankments consisting of lateral and vertical deformations, hydraulic conductivity, and field compaction data. Total stress on the foundation, changes in pore pressures, and the post-construction strength of the embankment were also investigated. Performance of the foundry sand section was compared to the performance of the two control sections. Mast (1997) concludes that, from a geotechnical perspective, foundry sands can be used as embankment fill in full-scale highway projects. Mast (1997) provides a wealth of information on the geotechnical properties of the foundry sand used in the study.

Javed and Lovell (1994) conducted a research project sponsored by InDOT on using foundry sand for structural fill. This project included characterization of seven foundry sands from green sand processes, two foundry sands from chemically bonded processes, and one from a shell molding process. The foundry sands from green sand processes had unit weights strongly dependent on water content due to the presence of bentonite. Maximum California Bearing Ratio (CBR) was observed at optimum water content. CBR values were comparable to values for compacted natural soils. Shear strength of the foundry sands was also measured using direct shear tests on dry samples. The shear strength parameters of foundry sands were found to be comparable to those for natural sands. Resilient modulus tests were performed to investigate the suitability of foundry sands in subbase applications. The resilient modulus test results showed that foundry sands have resilient moduli comparable to or even higher than those of soils typically used in subbase construction in Indiana. Similar results have been reported by Kleven (1998).

Specifications

Specifications for using foundry sands as fill materials are the same as those for typical granular backfills. These specifications vary depending on the specific use of the material, i.e. embankment, structural fill, roadway sub-base, and foundation sub-base. The specifications generally consist of compacting the material in layers of a maximum thickness (15-20 cm) to a certain percentage of the maximum dry unit weight (> 90%), and at least 15 kN/m^3. Compaction water contents should be close to optimum water content. Most specifications require a maximum liquid limit of 65%, and a plasticity index less than the liquid limit minus 30. Most foundry sands satisfy these requirements and are eligible to be considered as construction fill materials.

Recommendations

Foundry sand can be used as embankment material and fill material for typical construction applications such as sub-base, structural fill, etc. In all of these applications, foundry sand should be separated from other foundry by-products. Foundries should determine the particle size distribution, Atterberg limits, and compaction curves for their sand. Designers considering the material for beneficial reuse usually need such information. Permits for the beneficial reuse of foundry sand as a geotechnical fill are easily obtained when the foundry sand is to be used in covered areas or where the potential for leaching is minimal. For highway construction, most state regulations require that foundry sands to be encapsulated in a clay liner. Appropriate leach testing should be conducted per local environmental regulations to determine the environmental suitability.

PORTLAND CEMENT CONCRETE

Portland cement concrete, herein referred to as "concrete," consists of a mixture of approximately 30% sand, 50% gravel, 15% cement, and 55% water (Javed and Lovell 1994). Concrete can be either cast-in-place, or pre-cast into concrete products such as bricks, pipes, blocks, etc. The fine aggregate portion used in the production of concrete is of particular interest when foundry sands are beneficially re-used in concrete production.

Example Project

McIntyre et al. (1991) describe a study investigating the beneficial reuse of foundry sand in cast in-place concrete and concrete blocks. Several concrete mixes using foundry sand as a partial replacement for fine aggregate were developed and tested according to ASTM standards.

Concrete cylinders from several mixes were prepared for testing compressive strength after 28-days. The concrete mixes were designed using the American Concrete Institute (ACI) absolute volume method. This method involves using sieve analysis data, specific gravity, and bulk density to develop a mix design. Replacement of fine aggregates with foundry sand ranged from 15 to 20%. All mix designs used a water-cement ratio of 0.5, air content of 6%, and a target slump of 76 mm. Control mixes were prepared without foundry sand in the same proportions. Results of compression tests showed that cylinders made with foundry sand have the same strength as control cylinders when the fineness moduli required by ASTM C 25 are met.

Concrete blocks were cast using an industrial grade machine. Design mixes for the blocks were developed using the fineness modulus method, which is based on sieve analysis data. Three mixes were developed using 15, 30, and 45% replacement of fine aggregate with foundry sand. The 7-day and 28-day compressive strengths of the blocks were then determined. The compressive strength of the blocks decreased as the amount of foundry sand in the mix increased. However, the mixes containing 15% and 30% replacement produced blocks with strengths higher than the minimum strength (13.1 MPa) required by ASTM C 90 for Type N-1 blocks. All mixes containing 45% replacement of fine aggregate with foundry sand also produced blocks that met the ASTM C 90 strength criteria, except for one mix that had foundry sand containing no core sand.

The color of the blocks darkened as the quantity of foundry sand increased in the mix. Blocks using 15% replacement of foundry sand showed minimal darkness while blocks obtained using 30% replacement were considerably darker. Blocks using 45% replacement were considered too dark to be marketable.

McIntyre et al. (1991) recommend 15% replacement of fine aggregates with foundry sand in cast in-place concrete and in concrete blocks. This replacement ratio is likely to meet typical gradation and other specifications, such as fineness modulus.

Specifications

The specification for Portland cement concrete depends on the specific use of the concrete, such as bridge deck, pavement, blocks, etc. When foundry sands are used as a partial replacement for fine aggregates in the production of concrete the combined fine aggregates should have the gradation in Table 3, as recommended by ASTM C 33.

ASTM C 33 requires that the fineness modulus for the combined fine aggregate be 2.3 to 3.1, and not vary by 0.2 from the approved aggregate source. Most specifications also require that the percent loss in a sodium sulfate soundness test be less than 10%. In addition to these requirements, the concrete must meet project-specific strength requirements.

Table 3. Typical Gradation for Fine Aggregate in Concrete

Sieve Size	Percent Passing (%)
9.5 mm	100
4.75 mm (No. 4)	95-100
2.36 mm (No. 8)	80-100
1.18 mm (No. 16)	50-85
600 μm (No. 30)	25-60
300 μm (No. 50)	10-30
150 μm (No. 100)	2-10

Recommendations

Using foundry sand as a partial replacement of fine aggregate in Portland cement concrete is a large market that could use all of the foundry sand being produced in the U.S. Core sand (essentially clean sand with chemical binder) can replace as much as 45% of the fine aggregate in concrete. Green sand (sand with clay binder, typically up to 15%), however, can replace only 9 to 15% of the fine aggregate, depending on the amount of fines in the green sand. This percentage can be increased if the foundry sand is processed and the fines are removed.

ASPHALT CONCRETE

Asphalt concrete, also known as bituminous concrete, consists of a mixture of aggregates bound together by asphalt cement. The aggregates in this mixture are usually required to meet ASTM or AASHTO specifications. The most important properties of the aggregate are gradation, shape, and density because they affect the characteristics of asphalt concrete.

Example Project

Leidel et al. (1994) report that foundry sand has been used in asphalt concrete in Ontario, Canada since the early 1980's. Based on 15 years experience, Leidel et al. (1994) recommend that 15% of clean foundry sand can be used as a replacement of fine aggregates in asphalt concrete. Use of higher percentages of foundry sand without adding anti-stripping additives results in pavement deterioration shortly after construction.

Northland Bitulithic Ltd., an asphalt concrete producer in Ontario, has used foundry sand as a substitute for fine aggregate for 10 years. Two design mixes were developed based on their experience. One mix is for foundation layer and the other is for surface course of asphalt concrete pavement. A summary of the two mixes is in Table 4.

Table 4. Northland Bitulithic Limited Mix Design

Component	Foundation	Surface
Asphalt Cement (% of total mixture)	5.5%	5%
95 mm x #4 Coarse Aggregates (% of total aggregates)	40%	N/A
25 mm x #4 Coarse Aggregates (% of total aggregates)	55%	N/A
Foundry Sand	30%	22.5
Stone Sand	30%	22.5%

The General Motors foundry in St. Catherine, Ontario supplies the foundry sand used by Northland Bitulithic. The foundry sand is processed before use to remove metallic material and large particles.

Specifications

Specifications for asphalt concrete depend on the application. For foundry sands to be used as a partial replacement for fine aggregates, the combined fine aggregates should have the gradation summarized in Table 5.

Table 5. Typical Gradation for Fine Aggregates in Asphalt Concrete

Sieve Size	Total Percent Passing
9.5 mm	100
4.75 mm (No. 4)	95-100
2.36 mm (No. 8)	65-100
1.18 mm (No. 16)	40-85
600 μm (No. 30)	15-60
300 μm (No. 50)	7-40
150 μm (No. 100)	0-20
75 μm (No. 200)	0-10

The recommended gradation of fine aggregates in asphalt concrete varies depending on the type of asphalt and by location. The fines content (< No. 200 sieve) should be minimized to meet strength requirements and to avoid stripping. Most specifications also require that the percent loss in a sodium sulfate soundness test be less than 15%.

Recommendations

Using foundry sand as a partial replacement of fine aggregates in asphalt concrete represents a large market that could use all of the foundry sand produced in the North America. Foundry sand can replace as much as 15% of fine aggregates in asphalt concrete. This percentage can be increased if the foundry sand is processed and fines are removed. Foundries should supply the asphalt industry with data sheets describing their sand, including a particle size analysis and bentonite content.

PORTLAND CEMENT MANUFACTURING

Raw materials used in Portland cement manufacturing must contain the appropriate proportion of calcium oxide, silica, alumina, and iron oxide. Most of these necessary ingredients are usually contained in shale, dolomite, and limestone. When deficiencies in these ingredients are encountered in the rock normally used in cement production, additional silica and alumina are added. Since sand is a good source of silica. alumina, and iron oxides, it is often used in the production of Portland cement. Specifications for suitable sands include a minimum silica content of 80% (Leidel et al. 1994). Foundry by-products can also be used as a source of minerals. For example. foundry slag is a source of aluminum oxide and magnesia and foundry sand is a source of silica. Also, the clay fraction of foundry sand is a source of iron and aluminum oxides.

Example Projects
The American Foundrymen's Society (AFS) and the Construction Technology Laboratories (CTL), a subsidiary of the Portland Cement Association, co-sponsored the only research project on the suitability of foundry sands in Portland cement manufacturing (AFS 1991, AFS 1992, and AFS 1996). This study consisted of preparing raw cement mixes with and without foundry sand. X-ray fluorescence analysis showed that the foundry sand consisted of 88% silica, 5% alumina, 1% iron oxide, 0.19% sodium. and 0.25% potassium. The loss on ignition was 5%. The sodium and potassium levels were found acceptable.

Based on the x-ray fluorescence analysis, foundry sands were deemed an attractive alternative as a raw material for use in Portland cement manufacturing. Consequently, four samples of raw materials were prepared and proportioned with limestone, clay, sand, iron ore, and magnesium carbonate. Sand in three of the mixes was incrementally replaced by foundry sand in the following percentages: 0% (control sample), 4.4%, 8.9%, and 13.4% (complete replacement by foundry sand). All samples were then ground to pass the No. 200 sieve and introduced into a kiln at a temperature of 4800 °C. An oxide and free lime analysis performed on the clinkers showed no significant difference between clinkers obtained from the four different mixes. Four samples of cement were then made by grinding the clinkers and mixing them with gypsum. No significant difference was found between the compressive strength and the set time of the four cements.

Specifications
According to the Portland cement industry, foundry sand can be beneficially reused in the manufacturing of Portland cement when it possesses the following properties: (1) silica content $\geq$ 80%, (2) low alkali level, and (3) uniform particle size. In addition, large quantities of foundry sands must be available for it to be used by Portland cement manufacturers.

Recommendations
Foundry sand used in Portland cement manufacturing should be separated from other foundry by-products. Most Portland cement plants also require that core butts be ground to a uniform grain size.

AGRONOMIC SOIL AMMENDMENT

Soil mixtures used by the nursery and green house industries are typically sand and/or gravel mixed with peat, fertilizers, top soil, etc. Replacing natural sand with foundry sand in agronomic applications represents an excellent market for beneficial reuse of foundry sand. The presence of clay in foundry sand is beneficial because clay increases the capacity of soils to retain nutrients and water, therefore reduces the amount of additional nutrient required for plant growth (Voigt and Regan 1995).

Example Project
Dunkelberger and Regan (1997) investigated the use of foundry sands as agronomic amendments for horticultural production of ornamental flowers and shrubs. This study had two main objectives. One objective was to determine if foundry sands could be incorporated into growing mixes for the production of Orbit Red (a geranium) for the green house industry. The other objective was to investigate if foundry sands can be incorporated into growing mixes for the nursery industry.

Different growing mixes were prepared using foundry sand. Percent solids, air content, water content, and total porosity were determined for each mixture and compared to those of commercially available potting mixes. In addition, plant growth was evaluated for different types of foundry sands and compared to plant growth in commercial growing mixes. Dunkelberger and Regan (1997) conclude that plants, such as the petunia 'Spring Fever White' and the geranium 'Orbit Red' will grow in potting mixtures containing foundry sands. They also conclude that foundry sands can be successfully used in growth media or in manufacturing of potting soil for the production of nursery stock. In addition, foundry sands should be mixed with coarse amendment such as compost, to obtain mixtures with sufficient porosity.

Specifications
Neither the nursery industry nor the greenhouse industry has developed specifications or standards for soil amendments or soil mixtures (Voigt and Regan 1995). ASTM E 1598 is usually used to assess the toxicity of soil amendments obtained from mixing industrial by-products with conventional materials. This test should be performed on foundry sand-peat mixtures to demonstrate that foundry sand will not limit plant growth (Voigt and Regan 1995).

Recommendations
Dunkelberger and Regan (1997) recommend using a mixture of 50% manufactured growing mix and 50% foundry sand. Metals in foundry sands can be toxic to plants. Therefore the pH of the mix should be kept above 7. The nursery industry and the green house industry represent an excellent opportunity for small and large foundries to reuse their foundry sand. Such markets are typically local and are easily accessible.

LANDFILL LINERS AND COVERS

Most foundry sands being disposed in landfills are green sands. The primary components of green sand are silica and bentonite (Stephens and Kunes 1982). Thus, green sands are a sand-bentonite mixture, which makes them potentially useful as liner and cover materials (i.e., hydraulic barrier layers).

Example Project
Research is being conducted at the University of Wisconsin-Madison regarding the reuse of foundry sands as construction materials for barrier layers. The following issues are being addressed in this study:

- Which foundry sands are appropriate for constructing hydraulic barrier layers?

- What are the key properties a foundry sand must have for use as hydraulic barrier?
- What construction methods should be used with foundry sands?
- How durable are foundry sands when exposed to weathering distress or chemical permeation?

The project consists of laboratory and field studies (Abichou et al. 1998b, 1998c). Fifteen foundries from three states participated in the study. Index properties, compaction tests, and hydraulic conductivity tests were conducted on the foundry sands. The liquid limit (LL) ranged from 0 (non-plastic) to 29, the plasticity index (PI) ranged from non-plastic to 7, the bentonite content ranged from 0 to 16%, and the specific gravity varied from 2.53 to 2.73. The fines content typically ranged from 10 to 15%.

Hydraulic conductivities less than 1×10^{-7} cm/sec were achieved for nine foundry sands. These nine foundry sands had LL > 20 and bentonite content > 6%. All but one of these nine foundry sands was resistant to physical distress associated with freeze-thaw action and desiccation. The hydraulic conductivity after freeze-thaw cycling averaged 1.14 times the initial hydraulic conductivity (K_i) after 3 cycles. Similarly, the hydraulic conductivity after three desiccation cycles was 1.7 times the initial hydraulic conductivity, on average.

Short-term compatibility tests have shown that foundry sands are also resistant to chemical permeation by brine and typical MSW leachate, but not acidic solutions. Long-term compatibility testing is underway to confirm if the resistance to chemical permeation with brine and MSW leachate, observed in the short-term tests, still exists after long-term permeation.

Specifications
Abichou et al. (1998b) recommend that foundry sands used for constructing hydraulic barriers have bentonite content > 6% and LL > 20. Construction of landfill liners and covers using foundry sand should be performed using the same specifications employed for clay barriers. However, the hydraulic conductivity of foundry sands is not particularly sensitive to compaction water content and compactive effort. Relationships between hydraulic conductivity and compaction water content at different compactive effort should be obtained in the laboratory and used to develop an acceptable field compaction zone following the methods described in Daniel and Benson (1990).

SUMMARY AND CONCLUSIONS

The database contains information from 90 projects in 14 states and two locations in Canada. A broad variety of applications have been identified, and technical review of many of the projects was performed. The entire database is available electronically through Microsoft Access[®]. Based on the information gained through the development of this database, a master matrix that matches foundry by-products with beneficial reuse applications was developed. This matrix can be found in Abichou et al. (1998a).

Review of the database indicates that some applications (e.g., flowable fill, Portland cement production, and landfill liners and covers) are well-established. Specifications for these applications have been developed, and they appear to result in acceptable performance. However, some mix designs and specifications are proprietary, and thus are not available for widespread adoption. Other applications (backfills, subbase applications, asphalt) are far less developed, and general specifications are not yet available. However, for all materials there is a lack of field performance data and documentation. The absence of this information is an impediment to widespread beneficial reuse of foundry by-products. Another impediment to beneficial reuse is the lack of widespread dissemination and awareness of specifications and performance data for reuse applications. This database is intended to help remedy this shortcoming.

ACKNOWLEGMENT

Financial support for the study described in this paper was provided by the Wisconsin Cast Metals Association (WCMA) and the State of Wisconsin Recycling Market Development Board (RMDB). However, the findings described in this paper are solely those of the authors. Endorsement by WCMA or RMDB is not implied.

REFERENCES

Abichou, T., Benson, C., and Edil, T. (1998a), "Database on Beneficial Reuse of Foundry By-Products." Environmental Geotechnics Report 98-3, Dept. of Civil and Environmental Engineering, University of Wisconsin-Madison.

Abichou, T., Benson, C., and Edil, T. (1998b), "Beneficial Reuse of Foundry Sands in Construction of Hydraulic Barrier Layers." Environmental Geotechnics Report 98-2, Dept. of Civil and Environmental Engineering, University of Wisconsin-Madison.

Abichou, T., Benson, C., and Edil, T. (1998c), "Using Waste Foundry Sand for Hydraulic Barriers." In this GSP on Recycled Materials in Geotechnical Applications, GeoCongress-98, Boston.

American Foundrymen's Society, Inc. (1991), "Final (Phase I) Report on Alternate Utilization of Foundry Waste Sand." Illinois Department of Commerce and Community Affairs. Grant No. 90-82109.

American Foundrymen's Society, Inc. (1992), "Final (Phase II) Report on Alternate Utilization of Foundry Waste Sand." Illinois Department of Commerce and Community Affairs. Grant No. 90-82109.

American Foundrymen's Society, Inc. (1996), "Foundry Sand Beneficial Reuse Manual (Special Report)." Illinois Department of Commerce and Community Affairs. Grant No. 90-82109.

Ham, R. K., Boyle, W. W. and Kunes, T. P. (1981), "Leachability of foundry process Solid Wastes," Journal of Environmental Engineering, ASCE, No. 107, February 1981, pp. 155-70.

Daniel, D., and Benson, C. (1990), "Water Content-Density Criteria for Compacted Soil Liners," Journal of Geotechnical Engineering, ASCE, No. 116 (12), pp. 1811-1830.

Dunkelberger, J. A., and Regan, R. W. (1997), "Evaluation of Spent Foundry Sand as Growing Mix Amendments: Potential Beneficial Use Option," Proceedings of AFS 101st Casting Congress, Seatle, WA., April 1997.

Javed S., and Lovell C. W. (1994), "Use of Waste Foundry Sand in Civil Engineering." Transportation Research Record, No. 1486, Transportation Research Board, Washington D.C., 1995, pp. 109-113.

Kleven, J. R. (1998), "Mechanical Properties of Excess Foundry System Sand and Evaluation of its Use in Roadway Structural Fill." M. Sc. Thesis, University of Wisconsin-Madison. Madison, WI.

Leidel, D. S., Novakowski, M., and Pohlman, D. (1994), "Beneficial Sand Reuse: Making it Work." Modern Casting, August 1994, pp. 28-31.

Mast G. D. (1997), "Field Demonstration of a Highway Embankment Using Waste Foundry Sand," M. Sc. Thesis, Purdue University. Purdue, IN.

McIntyre, S.W., Rundman, K.B., Bailhood, C. R., Rush, P., Sandell, J., and Stillwell, B. (1991), "The Benefication and Reuse of Foundry Sand Residuals," Michigan Technological University, Houghton, Michigan.

Naik, T. R., Singh, S. S. (1997). "Permeability of Flowable Slury Materials Containing Foundry Sand and Fly Ash," Journal of Geotechnical and Geoenvironmental Engineering, ASCE, May 1997, pp. 446-452.

Ohio EPA. (1997), Beneficial Reuse of Foundry By-Products Files. Bureau of Surface Water.

Pennsylvania DOT. (1990), Specifications. Pub. 408.

Stephen, W. A., Kunes, T. P. (1984) "Cutting the Cost of Disposal Through Innovative and Constructive Uses of Foundry Wastes," AFS Transactions Vol. 81-84; pp. 697-708.

Voight, R. C. and Raymond W. R. (1995). "Beneficial Use of Foundry Residual Wastes," Project No. 94C. 1110C. The Environmental Research Institute. The Pennsylvania State University.

Wisconsin DOT (1997), "Flowable Fill as Bridge Abutment Backfill, Interim Report," Project B-59-166 and B-59-165.

Wisconsin DOT (1996), "Flowable Fill as Bridge Abutment Backfill, Construction report," Project B-59-166.

Wisconsin DOT (1996), "Flowable Fill as Bridge Abutment Backfill, Interim Report," Project B-59-166.

Wisconsin DNR (1997). Beneficial Reuse of Foundry By-Products Files. Bureau of Solid and Hazardous Waste.

Subject Index

Page number refers to the first page of paper

Author Index

Page number refers to the first page of paper